Springer-Verlag Berlin Heidelberg GmbH

Engineering

ONLINE LIBRARY

http://www.springer.de/engine/

G.W. Meetham · M.H. Van de Voorde

Materials for High Temperature Engineering Applications

With 67 Figures

 Springer

Geoffrey W. Meetham, M.B.E.
Consultant
formerly with Rolls-Royce
Derby UK

Prof. Dr. ir. Marcel H. Van de Voorde
Delft University of Technology
Faculty of Applied Sciences
Dept. of Materials Science and Technology
Rotterdamse Weg 137
2628 AL Delft
The Netherlands

formerly with
European Commission Joint Research Centre Petten (NL)
Brussels (B)
European Organisation for Nuclear Research (CERN)
Geneva (CH)

Cataloging-in-Publication Data applied for

Die Deutsche Bibliothek - CIP-Einheitsaufnahme
Meetham, Geoffrey W.:
Materials for high temperature engineering applications / G.W. Meetham; M.H. Van de Voorde.
Berlin; Heidelberg; NewYork; Barcelona; Hong Kong; London; Milan; Singapore; Tokyo:
Springer, 2000
(Engineering materials)

ISBN 978-3-642-63109-2   ISBN 978-3-642-56938-8 (eBook)
DOI 10.1007/978-3-642-56938-8

© Springer-Verlag Berlin Heidelberg 2000
Originally published by Springer-Verlag Berlin Heidelberg New York in 2000

Data conversion: MEDIO, Innovative Medien Service GmbH, Berlin
Cover-Design: de'blik, Berlin
Printed on acid-free paper   SPIN 10730039   62/3020 5 4 3 2 1 0

Preface

The need for materials with high temperature capability in industries such as electric power generation, transportation and materials production/processing has increased dramatically since the early 1900s. Process efficiency increases with operating temperature and early attempts to improve efficiency by raising temperatures were not always successful. Materials with the necessary capability were rarely available and the importance of high temperature materials in determining equipment performance and reliability was gradually appreciated. From the mid 1900s accelerating effort has been directed towards increasing the temperature capability of existing material systems and developing new material types. Understanding of material behaviour and control of component manufacture to ensure the desired behaviour have been key elements of these activities for all materials systems.

The importance of high temperature materials is reflected in the content of the major materials-related scientific and technology journals and in publications such as *"The Superalloys"* which discuss specific material types in detail. The intention here is to bring together, in one volume, the key features of all high temperature engineering materials. Chronological developments, current capabilities and future needs are discussed. The book covers materials ranging from those that have long been in commercial use such as high temperature steels, to materials which are in their early development stages such as the latest intermetallics. The book should be of benefit to materials engineers, mechanical engineers and production engineers. It should be relevant to engineers in training and engineers with industrial experience who need a review of the overall high temperature materials scene. It should also be relevant to materials specialists who require an easily accessible review of materials outside their specialisation. References are included to allow the reader to research any material system in greater depth.

The authors acknowledge help and advice from friends and former colleagues in the materials producer and user industries and in academia. They also acknowledge assistance with the preparation of the manuscript from former colleagues at the EC Joint Research Centre at Petten in The Netherlands.

G.W. Meetham MBE
M.H. Van de Voorde

Table of Contents

1 Introduction

1.1
Need for High Temperature Materials

Operation at high temperature is of fundamental importance to many major sectors of industry, including material production and processing, chemical engineering, power generation, transportation and aerospace. For the majority of processes, efficiency increases with increasing process temperature. Thus the ability to operate at high temperature is a crucial factor for industrial competitiveness. This ability is dependent on the availability of materials capable of withstanding the mechanical and environmental conditions of the particular operation. Typical operating temperatures for materials in industrial processes and machinery are illustrated in Fig. 1.

Maximum operating temperatures vary widely, depending on process, and the lives required also vary widely, ranging from minutes in some rocket applications to 100,000 hours in power generating plant. In all high temperature systems, it is essential to resist the mechanical and corrosive conditions imposed by the operating environment. Different applications have very different combinations of these conditions. Tungsten lamp filaments represent one end of the spectrum with environmental effects avoided by controlled atmosphere and operation being essentially stress free – the only stresses being thermally induced (start/stop) and direct (self-supporting weight). At the other end of the spectrum, the turbine blades of a military jet engine operate under high centrifugal stress with superimposed thermal stresses arising from rapid temperature changes during the flight cycle, in an environment that will be oxidizing at high temperature but may be corrosive at lower temperatures. At maximum power the hot combustion gas surrounding the blades is at a temperature above the melting temperature of the blade ma-

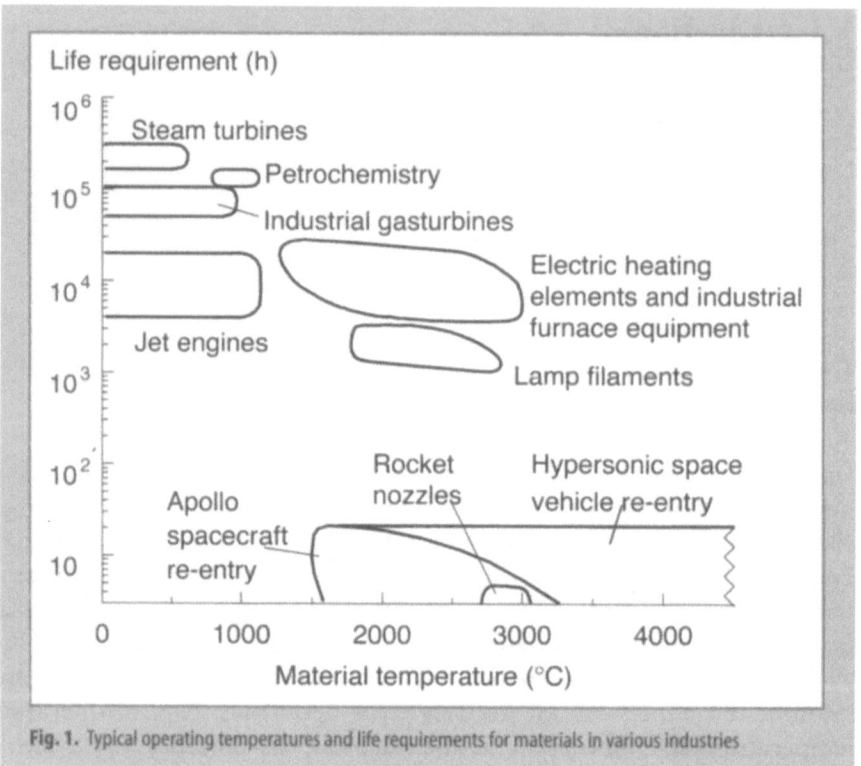

Fig. 1. Typical operating temperatures and life requirements for materials in various industries

terial and internal air-cooling is required to prevent the blade from melting.

The requirement to operate at progressively higher temperatures will remain an ongoing need for the foreseeable future. Many industries will benefit from increased operating temperatures. In electricity generation, the efficiency of ultra-supercritical pulverized coal power plant can be increased from the current 47% to 50% if steam parameters can be increased from 290-bar/580°C to 325-bar/625°C [1]. This will give major saving in fuel and consequent environmental benefits. The performance objectives of future military aircraft will require jet engines to have thrust/weight ratio increased from the current 12/1 to 20/1 together with reduced fuel consumption [2, 3]. Materials with higher temperature capability are essential if these, and many other, objectives are to be met.

1.2
High Temperature Materials

High temperature materials may be defined in several different ways, all of which are somewhat arbitrary. The least arbitrary definition might be based on the maximum use temperature as a proportion of the melting temperature, but even this depends on whether the application involves a significant stress level or is unstressed – Table 1. Most useful definitions are application based and involve some related specification of temperature. In this book, high temperature materials are taken to be those materials that are used specifically for their heat-resisting capabilities, such as strength or resistance to corrosion, at temperatures above 500°C. This allows the inclusion of titanium alloys and low alloy steels, which have provided substantial operational benefits because of increase in temperature capability resulting from long term material development programmes. It excludes the highest temperature aluminium alloys, which have only seen limited specific application. The stress/temperature capability of the major high-temperature materials systems is compared on a general basis in Fig. 2.

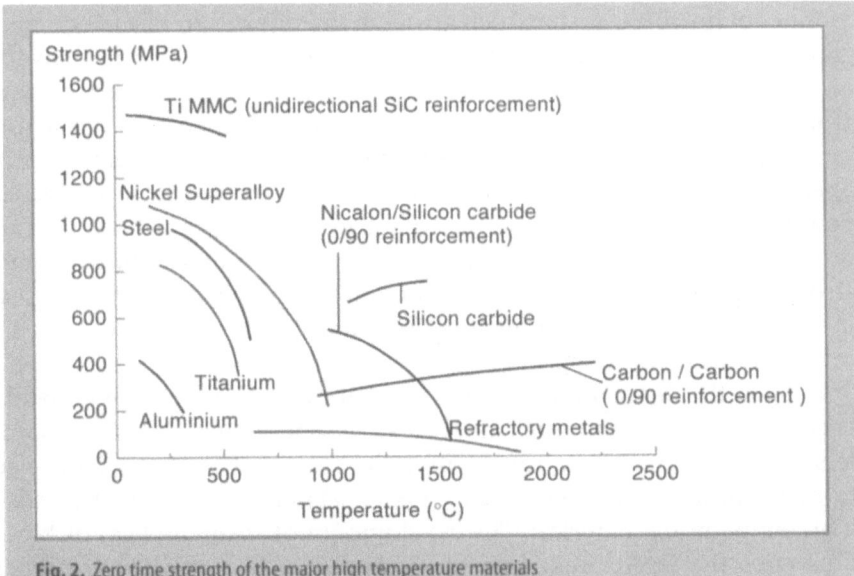

Fig. 2. Zero time strength of the major high temperature materials

Table 1. Maximum operating temperatures for various high temperature materials

Material	max. use temperature (ᵀC)	max. use temp. / melting temp.
Carbon steel	425*	0.27
Latest 12% Cr steel (HCM 12A)	650*	0.41
Nickel – 20% chromium alloy	400*	0.3
Single crystal nickel superalloy (CMSX4)	1050*	0.79
Oxidation res. nickel alloy (Brightray H)	1250#	0.9

* Stressed applications
# Unstressed applications

1.3
Historical Development of High Temperature Materials

The use of materials for structural purposes in high temperature situations dates back into pre-history, e.g. linings for cooking pits/hearths, early smelting of metals, etc. Since the industrial revolution, following a number of notable disasters such as steam boiler explosions for example, greater emphasis has been placed on developing materials specifically for high temperature applications. The need for materials with high temperature capability has accelerated from the turn of the century. The major industrial technologies which have been dependent on high temperature operation had their origins around that time – steam turbines, electric power generation, gas turbines, internal combustion engines and aircraft propulsion. The materials processing and petrochemical industries have also needed materials with high temperature capability. A chronological indication of the major industrial milestones and associated material milestones is given in Fig. 3.

Despite the fortunate discovery of material types which subsequently became very important, such as Brearley's accidental discovery of stainless steel in 1913, most advanced material systems have been developed to meet specific industrial needs. This is particularly the situation with high temperature materials. The development of steam and gas turbines illustrates the interdependence of industrial technology and materials development. The first successful steam turbines were designed by De

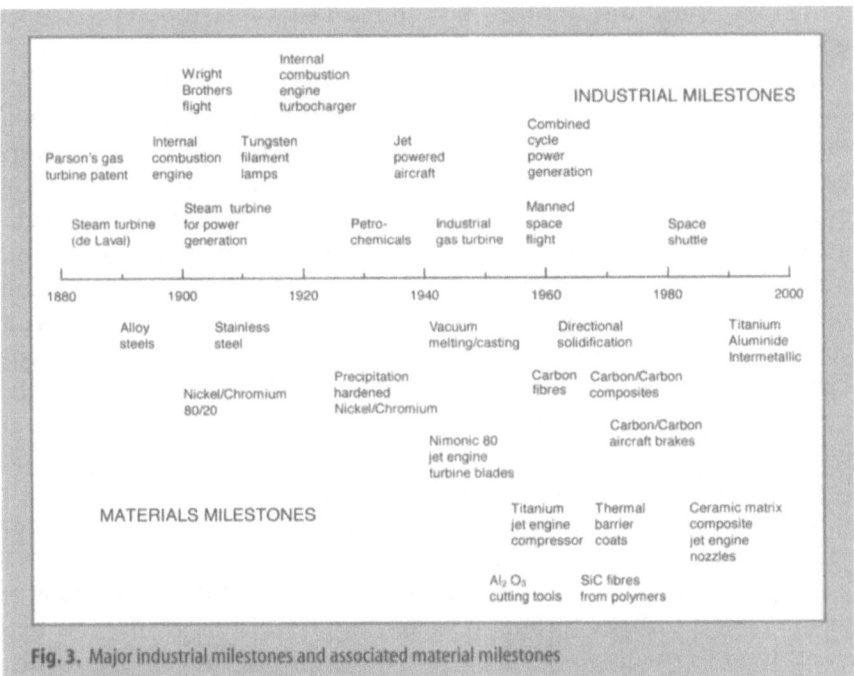

Fig. 3. Major industrial milestones and associated material milestones

Laval and Parsons in the 1880's. Parsons first demonstrated his turbine in 1897 in a 45-ton ship, the Turbinia, which achieved a speed some 10 knots higher than other ships at that time. Similar turbines are still extensively used in steam-powered electric generating plants and for propelling large ships. Wrought and cast iron had properties which were adequate for the earliest turbines where temperatures and stress levels were relatively low. It soon became apparent however, that it was necessary to operate at higher temperatures in order to increase plant thermal efficiency. This required the development of improved materials and low alloy steels with progressively increased temperature capability have been extensively used in steam raising plant and steam turbines, eventually to temperatures around 550°C. Applications requiring improved oxidation resistance made use of the benefits of higher chromium levels, as discovered by Brearley, in both ferritic and austenitic steels. Even today, 100 years after the first steam turbines, the need to increase the thermal efficiency of fossil fuel fired steam power plant is still limited by the creep strength of available materials. Further ongoing development of 9–12% chromium steels may allow operation of key components at temperatures up to 600°C.

Parsons established the fundamental principles of the modern gas turbine in his patent of 1884 [4]. The lack of materials with suitable high temperature properties for components such as turbine blades prevented the early development of the gas turbine for marine and land-based applications. World War II and the development of military jet powered aircraft led to the nickel superalloy materials, which were subsequently utilised in industrial gas turbines. The possibility of using the gas turbine for electric power generation was conceived in the early 1900's and was achieved in the years following World War II. In the 1960's, the concept of combined cycle operation increased the efficiency of electric power generation from 35% to 45%. In combined cycle operation the exhaust heat from a gas turbine is used to generate steam for a steam turbine. The efficiency of combined cycle plant has now been raised to more than 50% by increasing the turbine inlet temperature of the gas turbine [5].

The pioneering work by Von Ohain and Whittle [6] in 1937–1939 led to the concept of gas turbine powered aircraft. Initially, the low alloy steels used in steam power plant met the requirements of turbine discs but the operating conditions of turbine blades required a fundamentally better material [7]. Early trials with a precipitation hardened austenitic steel proved unsatisfactory. The excellent oxidation resistance of nickel chromium alloys had been recognized in the early 1900's and in 1929 it was observed that small amounts of aluminium and titanium added to a nickel-20% chromium alloy gave a significant increase in creep strength. The mechanism for this was not established until 1940 [8] and it provided the basis for the development of Nimonic 80, the first of the nickel superalloys.

The introduction of Nimonic 80 was accompanied by problems in manufacture, particularly the difficulty of producing a clean, oxide free material and the development of forging technology. Thus, even in the early days of nickel superalloys, there were indications of the importance of manufacturing processes and it is interesting to note that this has continued with many new superalloy developments dependent on new manufacturing processes such as vacuum melting and casting, directional solidification and powder metallurgy.

Early nickel superalloys have been widely used in the metal working industry for hot forging dies, extrusion dies and forging press tools because they are more resistant than tool steels to softening at high temperature. They have also been used for high temperature bolts, for exhaust manifolds and valves in internal combustion engines for high performance road vehicles.

The development of nickel superalloys with increased creep strength has involved compositional changes which have resulted in reduction of corrosion resistance, particularly in salt contaminated environments. In the early 1960's problems with turbine blade corrosion were experienced in aircraft operation involving such environments and protective coats were found to be necessary. Since that time it has become general practice to coat turbine aerofoils [7]. Since the late 1960's thermal barrier coats have become increasingly used to reduce metal temperatures in combustion chambers and turbine aerofoils. Thermal barrier coatings are now allowing higher temperature/more efficient operation or significantly extending the lives of components at their normal operating temperatures. These material technologies have also been applied in the progressive development of industrial gas turbines and in the application of aero-derived gas turbines in marine propulsion, with modifications which take account of the different operating environments.

The use of cutting tools in machining is an important element of manufacture. Early tools were made from carbon steel. In 1900 the benefits of tungsten and chromium in allowing faster machining speeds with associated higher tool tip temperatures were demonstrated and this led to the development of high-speed steel [9]. In the late 1920's cemented carbide tool tips were found to offer a further significant increase in machining speed. Al_2O_3 was considered as a tool material in the early 1930's because of its hardness retention at high temperature. It was not until the 1950's that strong, high quality Al_2O_3 tool bits were manufactured, but performance was found to be limited by low fracture toughness. In the late 1960's incorporation of TiC particles was found to increase toughness and this material has been used in mass production machining of cast iron. Subsequently further improvement in fracture toughness was achieved with the incorporation of silicon carbide whiskers in Al_2O_3 in the 1980's.

The concept of toughening by the incorporation of a reinforcing phase is particularly important in the most recent type of high temperature material – ceramic matrix composites [10]. Ceramics have been used in low stress-high temperature operations for many years. Silicon carbide has been used in furnace applications and electric heating elements. Exploration of the use of ceramics as structural components in reciprocating and

gas turbine engines began around 1970 [11]. This was motivated by predicted gains in performance including reduction in heat lost to the cooling system by insulating the combustion chamber of diesel engines and
avoiding the need to internally air cool components in gas turbine engines. Major ceramic engine programmes have been carried out in the
USA, Japan and Europe in attempts to capitalize on this potential. Lack
of defect tolerance and consequent unreliability of ceramics was a critical
barrier to achieving success. It was realized that defect tolerant ceramics
have been available since the 1960's in the form of carbon/carbon composites which were developed initially for rockets and re-entry vehicles.
Because of inherent oxidation problems with this material, surface protection is required at temperatures above about 400°C and reliable long
life coatings have not been produced. The major effort has thus been on
developing ceramic matrix composites with inherent oxidation resistance and SiC/SiC nozzle flaps have been incorporated in some military
aircraft since the early 1990's. Temperature capability is inadequate for
many applications however, and major research is ongoing worldwide
into high temperature non-metallic and intermetallic materials.

2 Design and Manufacture

For many years, design and manufacturing were considered as quite distinct and relatively unrelated stages in engineering plant construction. The designer specified the plant and the manufacturer was expected to make it. The benefit in designers and manufacturers working closely together to ensure that designs fully recognised the capabilities and limitations of manufacturing processes was gradually realized. It is only in relatively recent times, however, that the need for full integration of design and manufacture has been recognized. Currently industry is looking beyond computer-aided design (CAD), computer-aided manufacturing (CAM) and computer-aided engineering (CAE) to an objective of total computer integrated manufacturing (CIM) which, in addition to design and manufacturing, will also include manufacturing engineering such as process design, simulation, inspection and testing.

2.1
Plant Design and Material Selection

In the initial conceptual consideration of any new engineering plant type, the first step is to evaluate various possible designs against the predicted operating conditions, the associated technological and commercial requirements and any relevant national/international legislation. The fundamental objective is to achieve competitive performance with competitive life cycle costs. Maximum plant efficiency and minimum capital, operating and maintenance costs are key factors in meeting this objective. Avoiding plant failure is essential. The consequence of failure can vary from operational disruption with associated commercial problems, such as in production process plant for example, to hazarding human life, such as in the failure of public transport equipment. Reliability is determined by design and manufacture – the validity of the overall design concept, the ability of the various components of the plant to satisfy their design intent and their capability to meet specification in terms of properties and defects. Maintainability is mainly a function of design in terms of accessibility, inspectability and repairability.

Two complementary phases are commonly involved in the design process for engineering plant – the preliminary design phase and the detailed design phase. The preliminary design phase involves a study of possible basic concepts in order to select the design which best meets the overall product specification. Broad selection of material type is made in this phase with the emphasis on material cost, availability and prior use in related relevant environments. Simple models are used for structural calculations. As far as material properties are concerned, the emphasis is a failure-type data rather than on detailed characterization of material behaviour.

The subsequent detailed design phase requires accurate knowledge of the environment and loading to which the plant is subjected. The macroscopic response of the various components within the plant to their particular operating environment is determined. Finite element (FE) methods [1] are used in thermal and structural simulations to complete the detail design of the component. In this phase, detailed material specifications are considered and material selection is based on property data bases and understanding of material behaviour in relation to the environment and loads to which the plant will be subjected in service. It is essential that the materials data used in design adequately represent the properties in the particular manufactured form and also the scatter between different material batches and the scatter inherent in the manufacturing process.

In the early days it was general practice to design and life components on the basis of linear elastic behaviour. It was accepted that in certain regions of the component stress redistribution could occur because the yield strength was exceeded locally or because local creep deformation took place. The designer relied on the material having sufficient ductility to redistribute the peak stresses prior to the initiation of significant local damage. This approach has subsequently been shown to be unacceptable for highly loaded plant. Increasingly it is becoming essential to take account of inelastic behaviour in the analysis of macroscopic component behaviour by the use of refined FE analysis. This requires constitutive models for material behaviour to relate the stress at any position in the component to strain and temperature histories. Standard creep and plasticity models are available in FE programmes, but more complex models, which represent creep and plasticity together in time-dependent constitutive equations, are not yet available. In addition to establishing the macroscopic response of the component to the applied loads, it may also be necessary to characterise the microscopic response to the local temperature, stress and environment in critical areas of the component. This requires a thorough understanding of material behaviour. The exact sequence by which damage nucleates and develops is a key issue involving the effect of micro-scale damage on macro-scale behaviour, leading to the need for material behavioural models.

Much of the knowledge which is required in order to take account of inelastic behaviour is only now becoming available. Consequently many components in existing engineering plant have been designed with conservative safety factors, appropriate to the level of design capability at the time. Validated life extension of such components, together with life-extension refurbishment/repair techniques as appropriate, has become a significant factor in reducing cost of ownership in some major industries.

2.2
Component Manufacture

The manufacturing process is the means by which raw material is converted into the finished product. The success of the design process in terms of ability to deliver the required performance and achieve the predicted life depends on the manufacturing process producing, consistently and reliably, the component microstructure and maximum defect size and type implicit in the detailed design.

Developments in melting and mechanical working technologies have progressively improved the quality of the manufactured product, which can therefore be reflected in the design requirement. In steelmaking, electric arc furnaces were first used around 1920 and became the favoured melting route for early high temperature steels. The first improvement in material quality came with the introduction of vacuum degassing [2]. By 1950, reliable large scale pumping systems became available and it was found that, under certain circumstances, vacuum degassing could significantly reduce the non-metallic inclusion content of the steel. Since that time, some form of vacuum degassing became an integral part of steelmaking procedure. Even with vacuum degassing however property reproducibility, particularly in the transverse direction, was not adequate and further manufacturing improvements were necessary. From the mid 1950's vacuum arc remelting (VAR) was available as a remelting process. VAR had been developed primarily for the remelting of titanium which reacts with most refractories and with oxygen and nitrogen atmospheres. Remelting in vacuum into water cooled copper crucibles proved feasible. The vacuum in VAR is some three orders of magnitude better than that in vacuum degassing and produced major property improvements in steel [3]. Electroslag remelting (ESR) was introduced somewhat later as an alternative refining process. Like VAR it remelts into a water cooled copper crucible but air is excluded from the molten metal by a layer of molten slag rather than by vacuum. Both processes build up the ingot progressively and involve only a small amount of molten material at any one time. Compared with conventional ingot casting practice, this results in much reduced chemical segregation and an improved grain structure which facilitates subsequent forging operations. Both processes reduce non-metallic inclusions, particularly the larger particles. In VAR much of the oxide is removed as CO due to reaction with carbon, and in the ESR process oxides tend to be mechanically absorbed by the slag. In addition to these effects, because of the vacuum in the VAR process, volatile tramp impurity elements such as tin, lead, antimony etc. are removed to a large extent with benefits to hot working processes and to mechanical properties.

A further improvement can be achieved by producing the electrode for VAR or ESR by melting in vacuum rather than in air. This can be achieved by vacuum induction melting (VIM). The use of VIM significantly increases the overall cost however and it is not widely used for steelmaking other than for critical specialised applications such as in nuclear engineering. It is widely used for inherently more expensive materials such as nickel superalloys.

As noted, a major benefit from the remelt processes is improved microstructure to facilitate hot working. For critical high-temperature applications, the starting material for hammer or press forging operations is forged billet. This is to ensure that the starting material has been uniformly worked throughout its section. Billet forging is carried out on open press forges and for components such as shafts, the finish-forged shape is produced on open presses. Components such as turbine discs are forged in closed dies on hammers or presses [4]. Properties can be optimised for particular applications by the use of thermo-mechanical processing (TMP), in which forging parameters and heat treatment are selected and controlled to produce a predetermined microstructure in the finished component. Presses are better suited than hammers for thermo-mechanical-processing because the rate of deformation can be controlled throughout the forging stroke. However in press forging the metal is in contact with the die for a longer time than in hammer forging. There is thus a greater chilling effect and special insulating techniques are necessary to protect the surface of the workpiece. It is because of such considerations that hot die forging techniques have been developed. The dies in such processes are typically manufactured in nickel superalloys such as IN100. The dies are preheated to 750–1000°C by induction heating compared with 300–500°C in conventional forging. Normal strain rates are used. Hot die forging is not normally justified in the manufacture of steel components and is normally restricted to high cost materials such as nickel superalloys and titanium.

Various developments of hot die forging have been applied to nickel superalloys and titanium. In isothermal forging, the die and the workpiece are at the same temperature. Low rates of deformation are necessary in order to avoid adiabatic heating in the workpiece and thus maintain temperature control. Forging temperatures, depending on material, are in the temperature range 900–1150°C. At these temperatures it is necessary to use TZM molybdenum alloy dies which require that the forging process is carried out in vacuum. Superplastic forging is a specific example of isothermal forging in which the superplastic properties of the alloy are utilised [5]. Titanium alloys are superplastically forged commercially and some nickel superalloys can be superplastically forged.

There are many parts for which forging is not the appropriate manufacturing route, either because of geometric complexity or because the material with the required properties cannot be forged. Casting and fabrication are major manufacturing processes used in the production of such parts.

In casting, the particular process selected depends on the size, shape and complexity of the component and on the alloy being cast. Processes include air casting into sand moulds for the production of cast iron and steel parts, centrifugal casting of tubes in various materials and vacuum investment casting of nickel superalloys for gas turbine components. Good material quality and good foundry engineering are key requirements in the production of high quality castings. In the early days, material for casting was produced by air melting, with a final vacuum degassing stage subsequently being found to be beneficial in improving casting properties and reducing porosity and inclusions. Vacuum induction melted barstock is essential in the production of castings in nickel super-alloys, which contain reactive alloying elements. Good foundry engineering requires appropriate mould design and casting parameter selection to ensure optimum mould filling, solidification and microstructural control. Gating and feeding systems must be designed to provide progressive solidification from lighter to heavier sections thus minimising casting porosity [6]. Exothermic compounds, insulating materials and chills, as appropriate, may be incorporated to assist control of solidification. The investment casting process using ceramic shell moulds and cores is based on the ancient "lost wax" process. It is widely used, in vacuum chambers, in the production of castings in nickel superalloys and is capable of the required dimensional control necessary in net-shape castings [7].

Fabricated structures are commonly used in applications such as combustors and burners in furnace equipment, incinerators and gas turbines, exhaust systems, heat exchangers etc. The fabrications are typically based on fabricated and joined sheet metal, in materials ranging from steel to titanium and nickel alloys [8]. Cast and forged details are commonly incorporated. The sheet metal is shaped by various processes including pressing, deep drawing or stretch forming. Chemical machining may be used where other manufacturing processes may be impracticable or too costly, for example in producing local strengthening ribs or bosses. Joining processes are determined by component geometry, material and property requirements and include the various fusion welding, resistance welding and brazing techniques [9].

Manufacture of ceramic and composite parts is discussed in chapters 14 and 15 and coating application techniques in chapters 12 and 16.

2.3
Process Models

Traditionally the selection of process parameters used in the manufacturing process was based on experience and a costly and time consuming trial-and-error approach. The traditional approach to maintain quality within acceptable limits has involved increasingly intensive product inspection to remove unacceptable products from the production line. Such inspection is also costly and time consuming.

In the current environment with the emphasis on total quality, the trend is increasingly to couple together process design and process control. A properly designed and controlled process should produce only good products, thus minimising the need for product inspection. The control requires a scientific description of the manufacturing process and of the response of the material to the process – this is the process model. Wide use has been made of numerical techniques to analyse material behaviour, as discussed earlier but it is only relatively recently that such techniques have been applied to modelling the manufacturing process. Just as a design-related materials property data base is essential in the design process, so a process-related materials property data base is essential for the manufacturing process, including thermal, physical, mechanical data etc. together with constitutive models for materials and processing conditions, as illustrated in Fig. 4.

The VAR and ESR remelting processes produce controlled chemistry and microstructure through controlled solidification conditions as noted earlier. Further improvements are, however, continually being sought because the consequences of defects are potentially so great. Such process improvements involve modelling the process and improved control based on the modelling. [10,11,12]

The use of process modelling is becoming standard practice in the manufacture of key components in gas turbine engines [13]. In the case of discs, for example, a material behavioural model specifies the microstructure which is required to achieve the necessary mechanical properties throughout the disc. The process model includes:

- Microstructure/processing relationship (strain, strain rate, temperature)
- Simulation of material movement during the forging process by FE techniques

This allows the forging process to be designed – die shape sequence, forging temperature, speeds, etc. – to produce the microstructure specified in

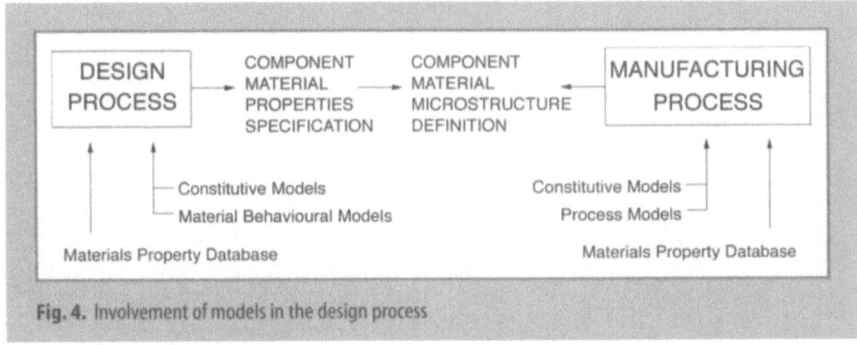

Fig. 4. Involvement of models in the design process

the material behavioural model. The sequence may be completed by modelling the subsequent heat treatment process. This allows the cooling rate from high temperature to be selected so that internal residual stresses are controlled within allowable limits. [14]

The design requirements in some of the most highly stressed components operating at high temperature are becoming so severe that manufacturing processes have been designed to produce the microstructure required for improvement of particular properties. In turbine rotor blades in the latest gas turbine engines for example, the directional solidification (DS) casting process produces a grain structure aligned from root to tip. In order to ensure maximum quality at minimum cost, it has been necessary to apply process modelling to such processes. [15,16] Finite element analysis is used to predict the solidification-time sequence of castings under actual foundry conditions. Local temperature gradients and solidification rates can be predicted and controlled in order to produce optimum microstructures and avoid defects such as microshrinkage and equi-axed grains within the DS microstructure, as illustrated in the schematic microstructural map Fig. 5.

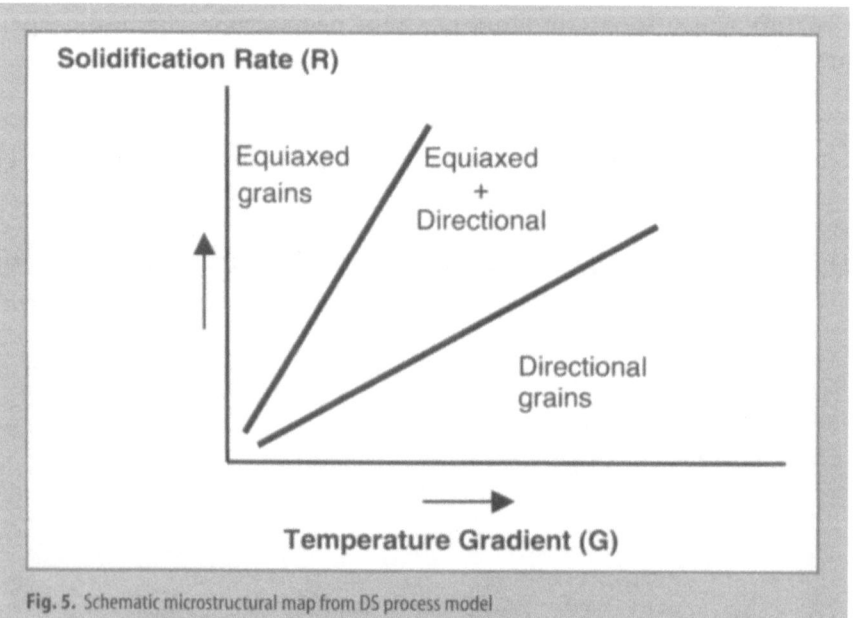

Fig. 5. Schematic microstructural map from DS process model

2.4
Component Life Extension

Life extension may involve several complementary activities:
- Material / component damage assessment
- Repair / rejuvenation as appropriate
- Life prediction and test validation
- Periodic re-inspection or on-line monitoring

The viability of any life extension possibility will depend on its relation to realistic overhaul/inspection schedules. The guaranteed minimum life extension must be well in excess of the minimum inspection period in service. High cost components are likely to be more appropriate candidates than low cost components.

Material damage assessment may involve a variety of techniques including:
- Microstructural examination of ex-service material (features such as precipitate coarsening, grain boundary changes e.g. creep cavitation, etc. assessed against control samples)
- Mechanical testing of material from critical regions of the component
- Non-destructive evaluation including X-ray, ultrasonics and magnetic techniques appropriate to the material/damage type under examination.

The material damage assessment and its extrapolation to remaining life potential and overall component integrity may indicate that the component has sufficient remaining life to justify further service without any refurbishment. Depending on the criticality of the component in service, it may be necessary to carry out supplementary component testing under simulated service conditions, to provide further validation evidence for an appropriate Certification Authority. For some components it may be appropriate to adopt a fracture mechanics approach to specify the timing of in-service inspections. A common philosophy is based on the maximum size of crack or defect which can be missed on NDE inspection and the subsequent growth rate of such a defect under service conditions [17].

The material damage assessment may indicate that some refurbishment or repair is necessary before further service usage is possible. As was noted earlier, such action may only be justified on high cost components because of the cost of developing and validating repair processes, the precise details of which may only be appropriate to one particular material/component geometry combination. Thus refurbishment and repair are becoming increasingly used on components of gas turbine engines, both large stationary turbines used for power generation [18] and smaller gas turbines used in aviation. The motivation is cost saving for the operator compared with replacing the service-run component with a new component, but the subsequent safe operation of refurbished parts is a crucial aspect. Refurbishment techniques include:

- Local dimensional correction to repair foreign object damage, particularly on the large blades of stationary gas turbines. Straightening and grinding may need to be supplemented by braze or weld deposition of new material, depending on the component base material and the location of damage in the component.
- Stripping and recoating. The turbine blades in both stationary and aero-gas turbines are commonly coated to minimize corrosion from the operating environment. It is not uncommon for the life of the coating to be reached while the base material still has a major portion of its life remaining. In such situations the coating is stripped, either locally or totally by chemical or mechanical processes. Recoating is then carried out, with the number of recoats permitted dependent on the remaining life of the base component and on the effect of recoating on the mechanical behaviour of the component [19].
- Regenerative heat treatment. The microstructural changes resulting from the combined effects of temperature, stress and time in service can lead to a progressive degradation in properties of the turbine blade material. In some forged turbine blades used in military engines, successful recovery has been achieved by re-heat treatment [20].
- Weld/braze material repair of cracks. Repair of cracked or eroded areas of gas turbine stator blades has become a common repair technique. The areas to be repaired are cleaned chemically or mechanically and are built up by material deposition or by joining of pre-shaped inserts [21]. Such techniques are not appropriate on the more highly stressed turbine rotor blades or on more recent stator blades of complex internal cooling configurations.

3 Requirements of High Temperature Materials

Selection of a material for reliable economic performance in a particular application must take account of many factors. Chemical resistance to the environment involved, mechanical behaviour and physical properties are of prime importance. In many applications involving high temperature, economic performance requires long life, up to several years in many industrial situations. During this time, the operation may involve many plant start-ups and shutdowns and not all of the operations involve the highest process temperatures. In some applications it is found that the low temperature stages of the operation may be the most damaging in corrosion terms. Thus, more than any other area where materials are used, high temperature applications may demand multi-material components if the design requirement is to be achieved. The material from which the component is made essentially meets the mechanical requirements of the operation and provides a modicum of environmental resistance, while a coating which is chemically and physically compatible with the base material may be necessary to provide the overall environmental resistance.

3.1
Environmental Resistance

High temperature corrosion problems can be complex. Depending on the environment, the types of high temperature corrosion encountered in industry can include oxidation, carburisation, sulphidation, nitridation, ash- and salt-deposit corrosion and molten salt corrosion. In most environments, oxygen activities are sufficiently high for oxidation to be involved in the corrosion process. Strongly oxidising environments involve high oxygen activities with excess oxygen. Reducing environments involve low oxygen activities, which are controlled by factors such as the CO/CO_2 or H_2/H_2O ratios in the environment. Reducing environments are generally more corrosive than oxidising environments because protective oxide scales form more slowly. In sulphidation, the severity of the environment is determined by the relative sulphur and oxygen activities and carburisation and nitridation behave in a similar way.

The basic factors involved in predicting high temperature oxidation and corrosion processes are thermodynamics and kinetics. Thermodynamics allows prediction of the corrosion product which is likely to form on a particular material in particular corrosion conditions. Kinetics are of concern in scale growth rates. However, whether the thermodynamically favoured scale will form in practice, and whether it will remain in place to provide protection against the environment, is determined by various other factors. These include contaminants in the material and the environment, multi-component diffusion, scale volatility, scale-substrate adherence and stresses developed in the scale. These stresses are influenced by factors such as coefficients of thermal expansion of scale and substrate and the mechanical elements of operation such as thermal cycling and creep.

Because of the complexity of theoretically predicting corrosion behaviour, many tests have been developed to evaluate corrosion resistance in simulated environments. These include, for example, the evaluation of hot corrosion behaviour of gas turbine blade materials by using molten salt crucible tests, electrochemical tests and burner rigs [1, 2]. Conditions can be developed in these tests which produce corrosion morphologies similar to those observed in service, but quantitative life prediction is nevertheless difficult. Such tests, together with knowledge of the real behaviour of related materials in service, are thus mainly used for material/coating selection purposes, with service trials providing final confirmation of behaviour.

3.1.1
Oxidation

Many industrial operations involve "clean" environments with few corrosive impurities. Examples include combustion atmospheres generated by clean fuels such as natural gas. In such environments oxidation is the high temperature reaction which develops protective scales on many materials [3]. Good protection is provided by Cr_2O_3 on stainless steel and by Al_2O_3 or Cr_2O_3 on nickel superalloys. The oxidation rate in these cases is usually parabolic – Fig. 6, curve a –, but factors which cause the oxide to

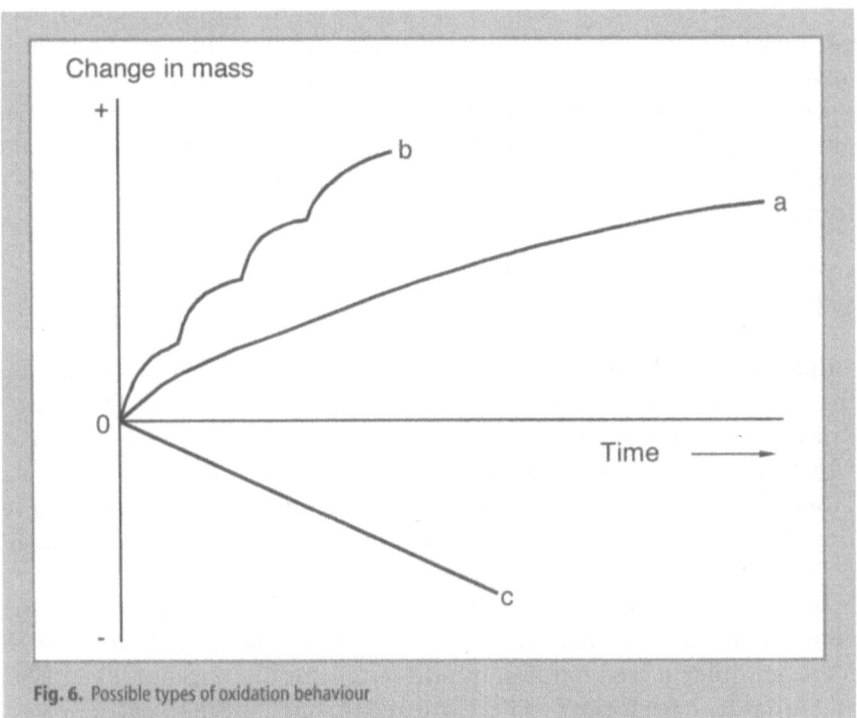

Fig. 6. Possible types of oxidation behaviour

spall – Fig. 6, curve b – render the protection ineffective. Some oxides, such as MoO_3 and WO_3 are volatile and offer no protection against oxidation – Fig. 6, curve c. Inability to develop reliable protective coatings has severely restricted the high temperature application of the refractory metals.

While oxidation most commonly occurs through the formation of a surface oxide layer, it can also occur internally within the microstructure of the alloy under conditions which promote the inward diffusion of oxidising species, often down microstructurally associated paths, e.g. grain boundaries. Various types of inter- and intra-granular oxidation can occur, sometimes well below the external surface of the alloy, leading to a substantially diminished load bearing section. In this event, the weight gain tests often conducted to indicate oxidation resistance would give a grossly misleading assessment of an alloy's performance.

Oxide ceramics are typically stable at high temperatures in oxygen or air. Non-oxide ceramics are not, although the oxide formed may provide some protection. Silicon carbide and molybdenum disilicide are protected in electric heating element applications, for example, by the layer of SiO_2 formed. Cracking of the SiO_2 because of phase changes during cooling can give rise to other problems. Sintered silicon nitride contains small amounts of metallic oxides added to aid densification on sintering and these can reduce oxidation resistance because they react with the SiO_2 formed, to produce complex surface glasses.

3.1.2
Sulphidation

Sulphidation – the formation of a sulphide surface scale – occurs when the sulphides of elements in the alloy are thermodynamically favoured rather than the oxides. It occurs primarily in gaseous environments which are oxidising with SO_2/SO_3 or reducing with H_2/H_2S gases, such as those found in combustion and coal gasification plants or petroleum refineries [4, 5]. Sulphidation is a very aggressive form of damage as the sulphide scales are rarely protective and form at rates several orders of magnitude higher than the corresponding oxides. Thus, in complex gaseous environments where it is possible that oxides or sulphides might form, complex mixed oxide/sulphide scales are often found. Even when stable oxides have formed, defects caused through mechanical damage or thermal cycling will allow sulphides to grow locally, undermining the oxide scale and leading to "breakaway" corrosion.

In both H_2/H_2S mixtures, typical of catalytic reforming and in SO_2 bearing environments, increasing chromium content has been found to be effective in reducing the extent of corrosion [6].

3.1.3
Salt- and Ash-Deposit Corrosion

It is not uncommon for deposits to form on component surfaces in industrial environments and the deposits can result in accelerated corrosion.

Hot Corrosion

Hot corrosion is the term applied to two specific types of corrosion experienced by gas turbines fired with fuels containing sulphur, and which usually have some exposure to marine environments. Sodium sulphate is formed from sodium compounds in the air ingested by the engine and sulphur from the combustion products of sulphur-containing fuel. This sodium sulphate is deposited on turbine stator and rotor blades. The protective Cr_2O_3 or Al_2O_3 scales are dissolved in the Na_2SO_4 due to fluxing mechanisms, resulting in accelerated "hot corrosion" [7, 8]. Two types of hot corrosion, with distinctive damage morphologies have been identified: –

Type I: The high temperature form shows extensive internal sulphidation under an alloy-depleted layer. It shows a peak in corrosion rate at a temperature which depends on alloy composition but is typically in the range 800–900°C.

Type II: The low temperature form shows little or no internal sulphidation and the corrosion rate peaks at around 700°C.

Deposit-induced Corrosion
In plant where deposits form readily, e.g. from the combustion of a contaminated fuel, enhanced forms of oxidation/sulphidation can occur through the formation of molten phases on the alloy surface. Two particular instances of this type of corrosion are superheater corrosion in coal fired boilers, especially when high chlorine coals are burned and boiler tube corrosion in waste incineration plant.

The highest tube metal temperatures in coal fired power plant occur in the superheater and reheater tubes and corrosion by the combustion gases together with deposits of combustion residues has been a limiting factor in maximum steam temperature. The deposits may contain sulphur, sodium and chlorine – the principal corrosive impurities from coal – together with the fly ash from the incombustible mineral matter in coal. The corrosion rate increases with temperature to a maximum and then decreases with increasing temperature in a similar way to Type II hot corrosion. The accelerated corrosion is related to the formation of a molten alkali metal iron sulphate [9].

Fireside corrosion in waste incinerators is increased by the impurities found in municipal or industrial wastes, which typically include sulphur, sodium, zinc, lead and chlorine, the latter being emphasised by the presence of any chlorine bearing plastics in the waste. Complex chloride and sulphate salts may be formed leading to sulphidation or chlorine attack. The successive formation and volatilization of metal chlorides at the metal/oxide interface, which condense and decompose to produce more oxide as they pass out through the scale, releases hydrogen chloride to repeat the cycle.

3.1.4
Carburisation

Carburisation may occur in equipment operating in low oxygen atmospheres containing significant amounts of carbon monoxide and hydrocarbons, including petrochemical processing, heat treatment equipment, carburising furnaces etc. Carbon from the atmosphere is absorbed onto the surface and forms carbides in the base material. These carbides reduce the chromium content in the base alloy and consequently the oxidation resistance, in addition to embrittling the surface layer of the alloy.

Fe-Ni-Cr alloys are widely used in carburising environments with nickel being beneficial in reducing the extent of carburisation. It has been reported that maximum carburisation resistance occurs with a $^{Ni}/_{Fe}$ ratio of 4, which may be related to carbon solubility and diffusivity [10].

Carburisation is generally a problem only at temperatures above about 800°C in industrial equipment. A very aggressive form of carburisation, known as metal dusting, can occur at lower temperatures, the extent peaking at around 650°C. The phenomenon appears to be related to the deposition of carbon from CO. Instances have been reported in waste heat boiler applications and in coal gasification plant [11], but the precise mechanisms involved in metal dusting have not been established.

Carbon also leads to a well-known form of damage in furnace lining refractories where the gaseous environment contains carbon monoxide. If the refractory materials used in the lining contain iron phases, the carbon monoxide diffusing into the refractory decomposes through the catalytic action of the iron to deposit pure carbon. Over a period of time the amount of carbon increases causing the surrounding material to fracture and disintegrate. This mechanism is known as carbon bursting.

3.2
Erosion

A moving fluid, usually containing solid particles, can remove material from a component by erosion. Erosion may be the only process operating, or it may be accentuated by corrosion. Erosion and corrosion frequently occur simultaneously – for example on water wall tubing in fluidised beds used for electric power generation. If the oxide scale is less resistant to high temperature erosion than the base material, enhanced corrosion will result as a consequence of the removal of the protective scale. If the oxide scale is more resistant to erosion than the base material, then it will provide a measure of erosion protection. Coatings are rarely found to provide effective resistance against erosion and appropriate design action is usually the only solution.

3.3
Wear

Wear takes place during the relative movement of parts in contact with one another. It has been a cause of component damage in many applications with very significant associated operational costs. In the mid 1960's, when knowledge of wear mechanisms in mechanical engineering applications was in its relative infancy, wear accounted for around 60% of the value of commercial aero-gas turbine components requiring repair or replacement [12]. Subsequent work established the necessary fundamental understanding of wear mechanisms, with wear processes occurring in the gas turbine engine being classified according to the predominant conditions of operation [13].

The type and severity of wear depend on many interacting factors including applied load, relative movement and temperature. Wear may be less severe at high temperature than at low temperature in some applications, because of the beneficial effects of oxides, particularly cobalt oxide. Poor oxide adherence however, can result in accelerated wear. Abrasive wear is caused by hard particles, including spalled oxides, between the rubbing surfaces, while adhesive wear is caused by micro-welding and subsequent shearing of asperities at contact surfaces.

Wear mechanisms can interact with processes such as corrosion, leading to enhanced metal loss rates, which can cause rapid failure in extreme cases.

3.4
Mechanical Behaviour

3.4.1
Zero Time Deformation

A basic preliminary design consideration is to limit applied stress levels to a specified proportion of the yield strength of steels or proof strength of non-ferrous metals. At stresses above the yield and proof strengths, zero-time deformation occurs. The major mechanism involved is slip, the movement of crystal planes relative to one another by dislocation movement [14]. Slip occurs more readily along preferred crystal planes and directions and thus the number of slip systems depends on the crystal structure. Slip is beneficial because, for example, it allows metals to be mechanically deformed to shape during manufacture and because it allows stress redistribution to occur in regions of local stress concentration in service, thus providing defect tolerance. Material strength is increased by making slip processes more difficult.

In contrast, few high temperature non-metallic materials are able to deform plastically in zero-time. Local stress concentrations can not thus be relieved and the materials are sensitive to the presence of defects. Other means must be found to introduce a measure of defect tolerance.

3.4.2
Creep

Creep normally involves slow, continuous real time deformation at temperature under stress. For metals, creep normally occurs at temperatures above 0.3/0.4 Tm, where Tm is the melting temperature in degrees absolute. Mechanisms that restrict zero-time deformation can also increase the temperature at which creep becomes a limiting factor. At relatively low temperature, creep is restricted by micro-structural features such as grain boundaries or precipitates restricting dislocation movement. At higher temperature dislocations can climb out of their blocked slip plane and continue the creep process. Diffusion is the controlling process in dislocation climb with the activation energies for creep and self-diffusion being similar for many crystalline solids [15]. This is illustrated by creep tests on iron at the α to γ phase transformation temperature of 910°C. Body centred cubic α iron tested just below 910°C had a creep rate around 200 times that of face centred cubic γ iron tested just above 910°C. This difference reflects the respective rates of self-diffusion. [16]

In the past, creep was considered to be the limiting factor in the design of many high temperature engineering components, with the requirement being to restrict creep deformation within design limits rather than to avoid creep rupture failure. In components such as turbine rotor blades, excessive creep would result in rubbing between blade and casing, while in high temperature bolts creep would lead to stress relaxation with time and the consequent need for re-torquing. Superheater and reheater

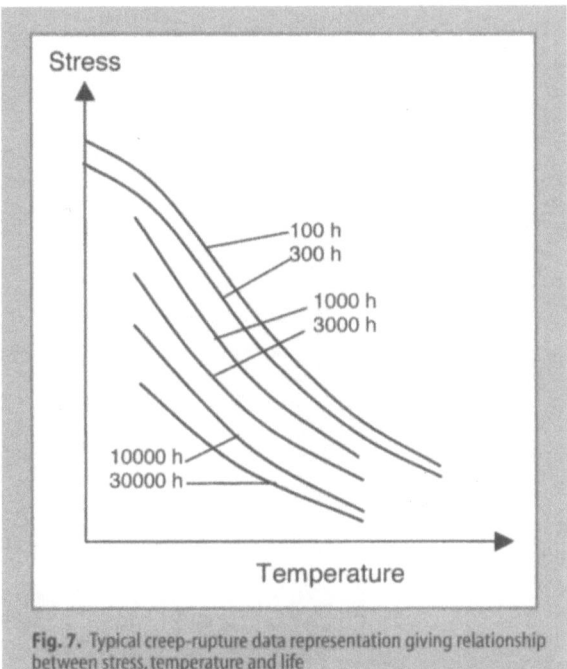

Fig. 7. Typical creep-rupture data representation giving relationship between stress, temperature and life

tubes in ultra-supercritical pulverised coal power generation plants have been creep limited and have required improved creep strength materials in order to operate at higher temperatures. In components such as the superheater outlet header and the latest air-cooled gas turbine blades, however, the major design consideration involves thermal stresses leading to complex creep-fatigue interactions. Creep considerations alone are becoming less common in high temperature plant.

Preliminary material comparisons are commonly made on the basis of stress to produce rupture in specified times over a temperature range – Fig. 7, assisted by derived empirical relationships such as that due to Larson and Miller. Design data is commonly presented in the form illustrated in Fig. 8 for various temperatures of interest.

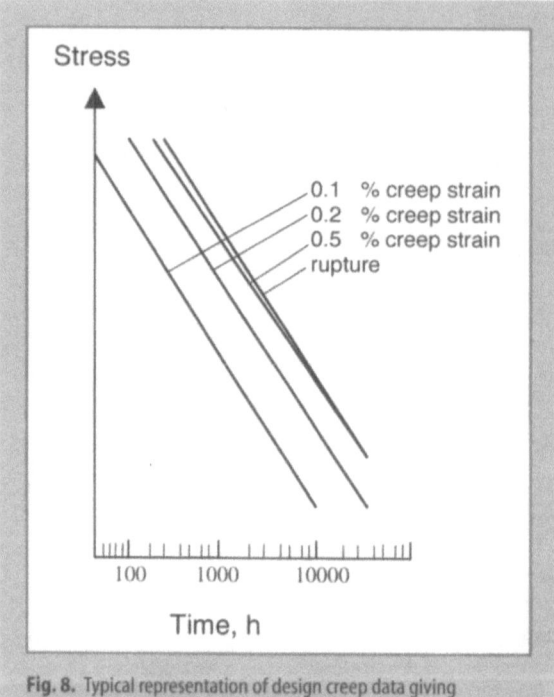

Fig. 8. Typical representation of design creep data giving minimum data for stress, time, creep strain at a particular temperature

3.4.3
Mechanical Fatigue

Mechanical fatigue failure can occur at stresses well below the tensile strength, when a material is subjected to repeated stress or strain cycles. Examples of cyclic loading in engineering plant include alternating bending loads on shafts and vibrational stresses on turbine blades. Fatigue is commonly classified as high-cycle fatigue (HCF) and low cycle fatigue (LCF) with the former referring to failures occurring above about 10^4 cycles. There is, however, no phenomenological difference between the two types of fatigue.

HCF data are usually presented as S-N curves, based on stress cycled axially loaded or bend tests. Data for a nickel superalloy, IN 625, are shown in Fig. 9, giving the relationship between stress range and cyclic life.

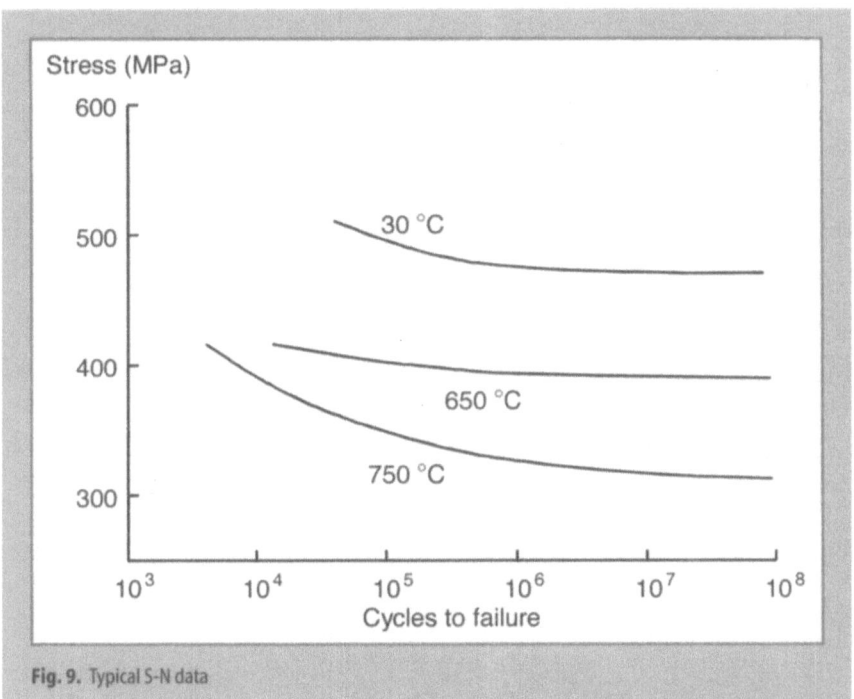

Fig. 9. Typical S-N data

Stress-concentrating features significantly reduce fatigue strength. These may be design features such as sharp changes in shaft diameter, poorly designed corners or holes or they may result from poor manufacturing practice such as machining marks, casting defects, undercut welds etc. Fatigue strength is very sensitive to surface condition. Techniques which increase surface strength, such as surface hardening and techniques which introduce residual compressive stresses, improve fatigue strength. Steam turbine blades are commonly shot peened for this reason.

Low cycle fatigue is a major consideration in much advanced engineering plant. It generally involves significant cyclic plasticity and failure occurs in a smaller number of cycles than in HCF. An extended period under load is commonly involved in LCF. LCF tests are carried out in basically the same manner as HCF tests except that the strain range is kept constant. The variation of stresses with strains results in a hysteresis loop, illustrated in Fig. 10 in which $\Delta\varepsilon$ is the total strain range and $\Delta\sigma$ is the total stress range.

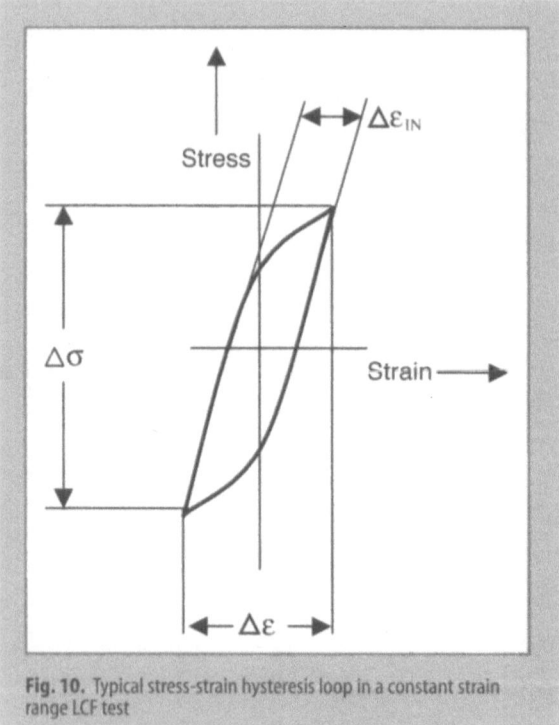

Fig. 10. Typical stress-strain hysteresis loop in a constant strain range LCF test

During the course of the test, the stress range does not remain constant, but either increases or decreases, depending on whether the material experiences cyclic strain softening or cyclic strain hardening. Eventually the stress range reaches an approximately steady value. Softened annealed materials typically undergo strain hardening while hardened materials typically strain soften.

Fatigue life is governed by $\Delta\varepsilon_{in}$, the inelastic strain range indicated in Fig. 10, according to the Manson-Coffin relationship $\Delta\varepsilon_{in} = aN_f^b$, where a and b are constants [17]. Whereas in HCF most of the life is spent in crack initiation, in LCF most of the life is spent in crack propagation. Thus, crack propagation rates are a major consideration in improving LCF performance, and microstructure has been found to have a significant influence on crack growth rates in nickel superalloys [18].

3.4.4
Thermo-Mechanical Fatigue

Many components which operate at high temperature experience transient temperature gradients during operation. Each transient temperature gradient generates a thermally induced stress and repeated transients generate cyclic stresses. Combination of thermal stresses with periods of dwell on-load at temperature results in creep fatigue interactions commonly referred to as thermo-mechanical fatigue (TMF). Several factors affect the severity of the TMF. The large component size in steam turbine rotors and casings results in large temperature gradients even under slow startup conditions. Physical properties of materials have a significant influence. Gas turbine blades present a particularly complex situation [19, 20]. Military aircraft missions in particular involve frequent engine accelerations and decelerations in any flight cycle, producing extreme thermal gradients and complex load cycles. Industrial gas turbines, in comparison, experience less severe thermal cycling in normal operation but emergency shutdowns generate high thermal stresses. In addition to the effect of operating conditions, different thermal and mechanical loading cycles are experienced at different aerofoil locations. Some locations will experience in-phase load-temperature cycling – Fig. 11a – some experience out-of-phase cycling – Fig. 11b – although in practice actual cycles may vary considerably from these idealized examples.

Out of phase cycling is generally the most detrimental because stress relaxation at high cycle temperature can increase tensile stresses at lower cycle temperatures and mean cycle stresses. Such a cycle type occurs at start up at aerofoil leading and trailing edges.

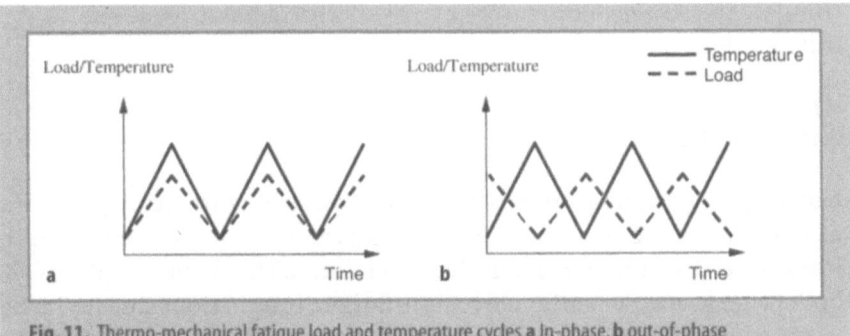

Fig. 11. Thermo-mechanical fatigue load and temperature cycles **a** In-phase, **b** out-of-phase

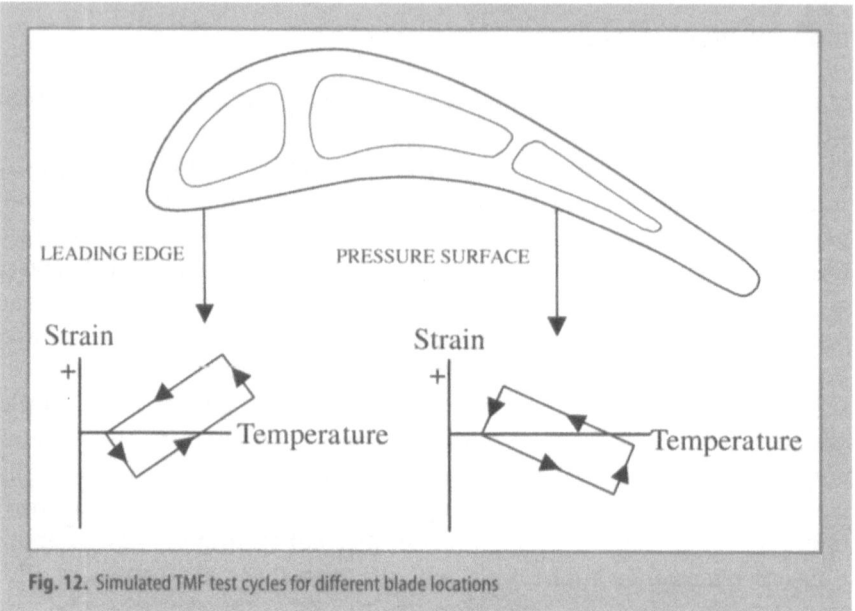

Fig. 12. Simulated TMF test cycles for different blade locations

Early thermal fatigue testing involved thermal cycling of particular shaped test pieces such as aerofoils or "Glenny" wedge test pieces. At best, such tests could only be used to compare relative materials performance and had no relevance to component lifing under thermal fatigue conditions. Current TMF tests are aimed at quantifying the relationship between temperature, strain and the corresponding stress and life of test samples [21,22]. Temperature and strain changes are applied to the test sample independently and strain and temperature are varied with time to simulate the strain-temperature cycles at critical blade locations, as illustrated in Fig. 12.

3.4.5
Corrosion-Fatigue

Not only can different sources of mechanical stress interact to give failure of components, but interactions with corrosion mechanisms resulting from the environment can also lead to failure. The furnace walls of shell boilers are known to suffer corrosion-fatigue interactions due to the high temperature corrosive environment and the cyclic stressing of the fireside surfaces due to the cyclic operation of the plant. In most cases, these cracks are detected during boiler inspections and remedial action taken before in-service failures can occur.

3.5
Physical Properties

Density
Density is of particular importance in high-speed rotating machinery, such as gas turbines, where centrifugal stress is a major element of overall stressing. Materials with high specific strength are preferred in such applications. Density also influences the natural frequency of vibration of components and may consequently influence fatigue behaviour.

Thermal Expansion Coefficient
Thermal expansion is important in controlling clearances and consequent stresses developed as a result of temperature changes in any assembly of component parts of different materials.

Thermal expansion coefficient is probably the most important physical property in determining the thermal shock resistance of engineering ceramics.

Thermal Conductivity
Together with thermal expansion, thermal conductivity plays a major role in determining the thermal stresses developed in a component as a result of temperature change. Low alloy steel was replaced with austenitic steel in thick section components in the first boilers operating with super-critical steam conditions in power generation in the early 1960's. Performance under base load conditions was good, but fatigue failures were experienced when operation was changed to load cycling, because of the additional thermal stresses developed. Such problems are less severe with ferritic/martensitic steels because of their lower coefficient of thermal expansion and higher thermal conductivity.

Conversely, some materials are used specifically for their low thermal conductivity. Where insulation is required to protect an underlying component from high temperature, low conductivity materials are exploited. Such materials are insulating refractories and thermal barrier coatings.

Young's Modulus

Young's modulus (E) also influences thermal stresses. Directionally solidified nickel superalloy turbine blades have superior thermal fatigue resistance compared to the earlier conventionally cast blades. A major factor in this improvement is that the natural crystal growth direction has a significantly lower E than other crystal directions.

Hardness

The hardness of a material is related to the material characteristics which give stiffness and strength. Hardness is an important characteristic of a material, as it contributes to resistance to erosion/wear processes. At high temperature, however, engineering alloys become 'softer' and so ceramics are often used to give wear resistance. The oxide layers forming on engineering alloys are also harder than the underlying alloy and will give some measure of wear resistance where they remain intact.

4 Increasing Temperature Capability

Fundamental requirements in increasing temperature capability include increase in strength at high temperature and increase in environmental resistance. However, the measures adopted to increase strength may be detrimental in terms of environmental resistance. Thus in many cases the overall design requirement cannot be satisfied by a single material. Coatings are commonly required. Depending on the complexity of component duty it may be necessary, in extreme cases, to apply multiple coatings to protect against different environmental factors, for example, corrosion and wear.

4.1
Metallic Materials

The various complementary mechanisms which strengthen metallic materials and increase temperature capability include solid solution strengthening, precipitation strengthening and dispersion strengthening. They all operate by making dislocation movement more difficult. Solid solution strengthening is applicable to all base metals. Precipitation hardening is a powerful strengthening mechanism but is limited to certain alloy types. Dispersion strengthening using fine dispersions of stable particles can be effective in developing strength at very high temperatures.

Grain boundaries play a very significant role in relation to material strength. Depending on temperature, they can be beneficial and play a role in increasing strength or can be detrimental and reduce strength. Thus grain size may be controlled within specified limits, the behaviour of grain boundaries may be modified and the boundaries themselves may be eliminated through the use of components in the form of single crystals.

4.1.1
Solid Solution Strengthening

When atoms of one metal are substituted into the crystal lattice of another metal, internal strains are generated and result in strengthening. The extent of the strengthening produced depends on the atoms involved. Atoms with similar crystal structures and lattice parameters will have high mutual solubility and will generate relatively little strengthening. Atoms of different size may have limited solubility but could generate significant strengthening.

The effect of various alloying elements on the proof stress of γ iron is shown in Fig. 13, and as is the case with α iron, the interstitial atoms N and C are more effective than the substitutional atoms.

Tungsten and molybdenum have long been recognized as potent solid solution strengthening elements in nickel superalloys. Recently, rhenium has been found to be a particularly effective element, partitioning mainly to the matrix, reducing diffusion rates and therefore retarding coarsening of the γ′ precipitate. It also forms short range ordering with very small

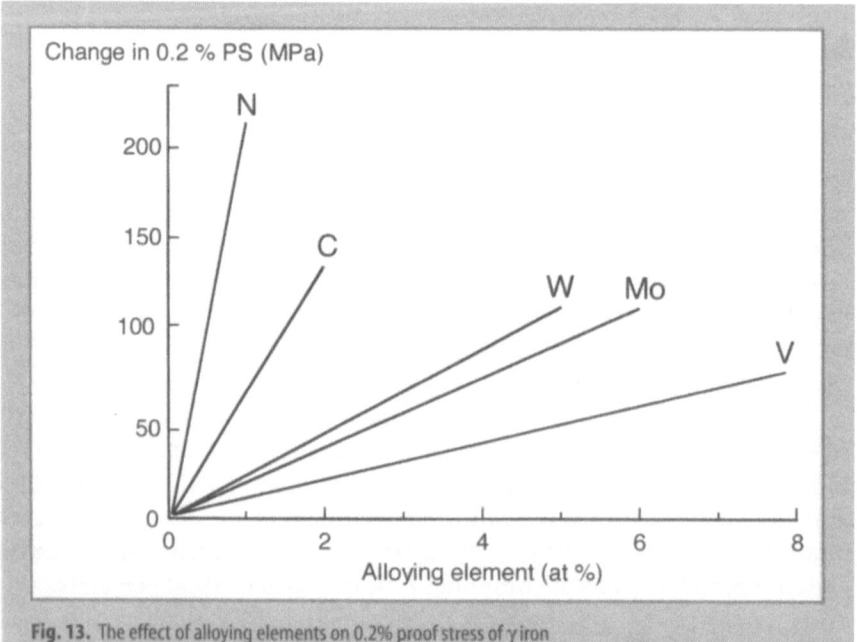

Fig. 13. The effect of alloying elements on 0.2% proof stress of γ iron

clusters of atoms in the matrix, which act as effective obstacles to dislocation movement.

4.1.2
Precipitation Strengthening

In many alloy systems it is possible to dissolve a second phase in the matrix by heating above the solvus temperature, retain the phase in solution by rapid cooling and reform the phase in controlled morphology by reheating to a temperature below the solvus. This is illustrated schematically for Ni-Al alloys in Fig. 14.

The precipitates impede dislocation movement. When the dislocations reach the precipitates, they must by-pass them by climbing or looping or must cut through them. Both mechanisms operate depending on precipitate size as illustrated schematically in Fig. 15 and there is an optimum precipitate size for maximum strengthening in different alloy systems. [1]

The strengthening produced depends on the compatibility between precipitate and matrix in crystal structure and lattice parameter and on the volume fraction of the precipitate and its size. Relatively small differences in lattice parameter allow coherence, with the structures of precipitate and matrix matching at the interface. The local distortion produced

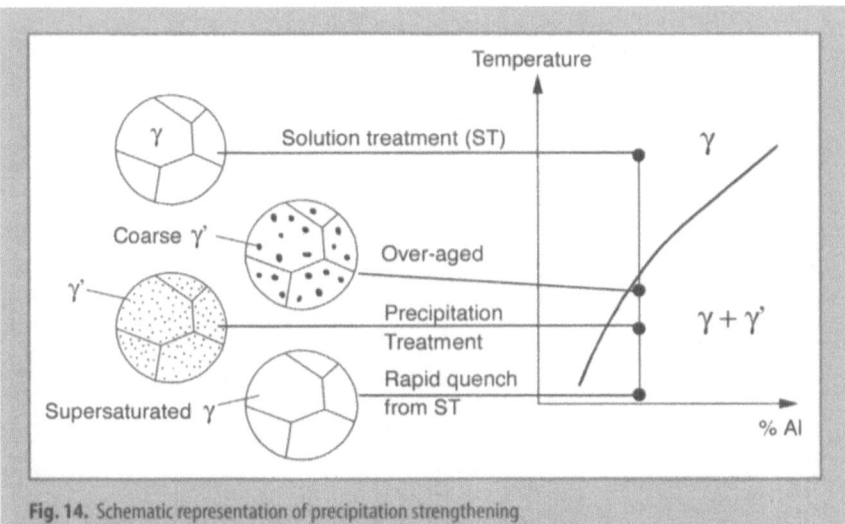

Fig. 14. Schematic representation of precipitation strengthening

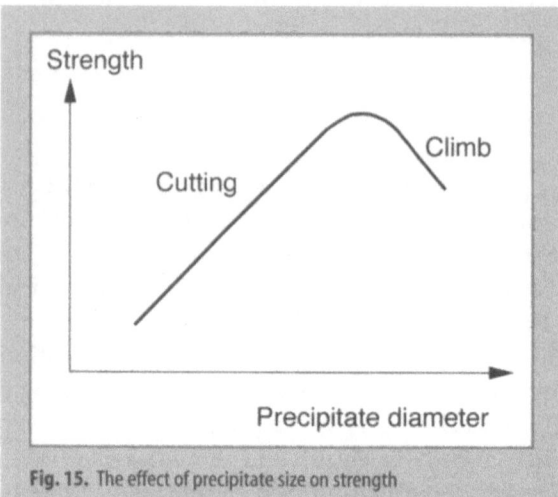

Fig. 15. The effect of precipitate size on strength

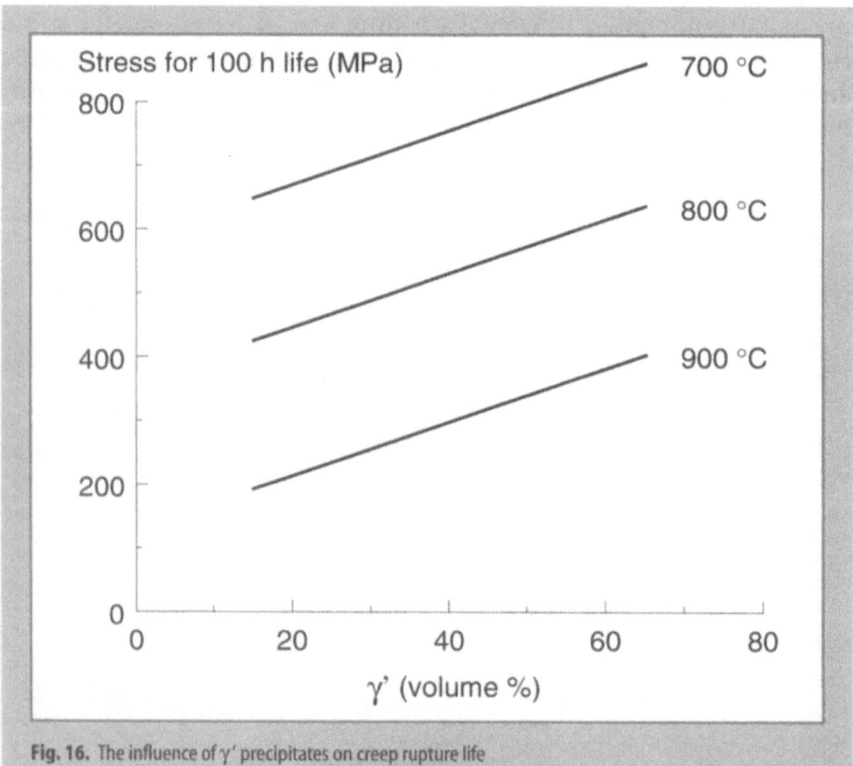

Fig. 16. The influence of γ' precipitates on creep rupture life

depends on the extent of mismatch in lattice parameters, with coherent precipitates capable of providing significant strengthening. Precipitates with higher mismatches are less effective in strengthening but this may be offset by increasing the volume fraction of precipitate.

Precipitation hardening is very effective in two of the major high temperature materials types-steels and nickel superalloys. Precipitation occurs during the tempering of martensite in low alloy steels, with Mo_2C, V_4C_3 and TiC having coherent interfaces on tempering in the temperature range 450-600°C [2]. Precipitation hardening in 12% chromium steels involves a chromium – rich M_2X phase containing carbon and nitrogen. Precipitation hardening in nickel superalloys is based on the intermetallic Ni_3Al phase (γ'), although titanium, niobium, tantalum and hafnium can replace some of the aluminum, modifying the behaviour [3]. The influence of amount of γ' precipitate on the creep rupture life for a range of nickel superalloys is illustrated in Fig. 16.

4.1.3
Dispersion Strengthening

The strengthening precipitates in precipitation hardening systems progressively coarsen and re-dissolve at higher temperatures. This limits their temperature capability. An effective way to produce strengthening at very high temperatures is to incorporate a fine, uniform distribution of dispersoid particles that are essentially insoluble in the matrix, such as Al_2O_3 in aluminum and ThO_2 in nickel. The dispersoid phase is not coherent with the matrix and so the strengthening effect is not as pronounced as with precipitation hardening but it is maintained to higher temperatures.

4.1.4
Grain Size and Grain Boundary Effects

A grain boundary is a region of mismatch between the lattices of adjacent grains. The effect of the grain boundary on properties varies with temperature. At temperatures up to around 50% of the melting temperature the boundary impedes dislocation movement and thus provides a strengthening mechanism. At higher temperatures, diffusion becomes increasingly important and is much more rapid in the grain boundaries than within the grains. Grain boundaries are therefore sources of weakness in high temperature creep processes. Depending on the service temperature of the component, the grain size and shape can be controlled, and the grain boundary structure can be modified, to optimize properties.

At moderate temperature, the yield strength of the material depends on grain size according to the Hall-Petch relation. Grain size control can be achieved in the manufacture of discs for gas turbine engines by thermo-mechanical processing (TMP) in which the temperature and strain rate of the forging process are controlled within appropriate limits. The effect of forging conditions and grain size on the proof and fatigue strength of INCO 901, a nickel-iron base superalloy [4] are illustrated in Table 2. This effect has been used to improve component fatigue life. However, incorporation of the growth behaviour of short cracks originating at critical defects, into component lifing calculations, has suggested that a structure consisting of relatively coarse grains separated by a necklace of fine grains may be optimum for highly stressed turbine discs [5]. Forging conditions can be controlled to produce such a microstructure.

As far as grain boundary weakness at high temperature is concerned, it is standard practice that nickel superalloys usually contain sufficient carbon content to introduce carbides in the grain boundaries to inhibit grain boundary sliding. These carbides must be controlled carefully because continuous carbides will result in brittleness. It has also become standard practice for nickel superalloys to contain other elements which segregate to grain boundaries and reduce diffusion in these regions – Table 3 – thus producing significant strengthening.

High temperature creep strength can be increased by increasing grain size to decrease grain boundary area and this approach has been used in

Table 2. Optimization of strength of Inco 901 by grain size control

	Conventional forge + heat treat 1090 + 775+705 ºC	Modified Forge +HT980+720+650 ºC
Grain Size	ASTM 2	ASTM 6
0,2% PS (MPa) 575 ºC	752	839
LCF (0-850 MPa) 500 ºC	$>10^4$ cycles	$>3 \times 10^4$ cycles

Table 3. The effect of Boron and Zirconium on the creep of Udimet 500

	UDIMET 500 Base Alloy	UDIMET 500 - Base Alloy +0.19%Zr +0.009% B
Creep life (h) 175 MPa 870 °C	50	650
Rupture ductility (%)	2	14

some situations. However, considerations such as thermal fatigue resistance impose a maximum on the grain size which is acceptable in components such as gas turbine blades.

Grain size control has been extended to include grain shape in directional solidification – the latest manufacturing technique for gas turbine blade manufacture [6]. In conventional castings, creep cracks initiate at grain boundaries which are normal to the direction of applied stress. Directional solidification produces a microstructure in which the grains are all aligned from blade root to tip so that there are no grain boundaries normal to the major applied stress. Creep life is thus extended. The remaining grain boundaries have small elements which are near normal to the applied stress and can thus initiate cracks as illustrated schematically in Fig. 17. Single crystal castings, with no grain boundaries at all, have higher creep life.

4.1.5
Environmental Resistance

As was noted in earlier, high temperature applications may demand multi-material components in order to meet the design requirement. The environmental resistance of the base material is normally optimised, consistent with its mechanical requirements, and if necessary a coating is applied to provide additional environmental resistance.

Experience with various materials in industrial environments provided early evidence on the effect of alloy composition on environmental resistance. Supplemented by fundamental research and simulated test procedures, this has led to the development of materials and coatings with significantly enhanced environmental resistance.

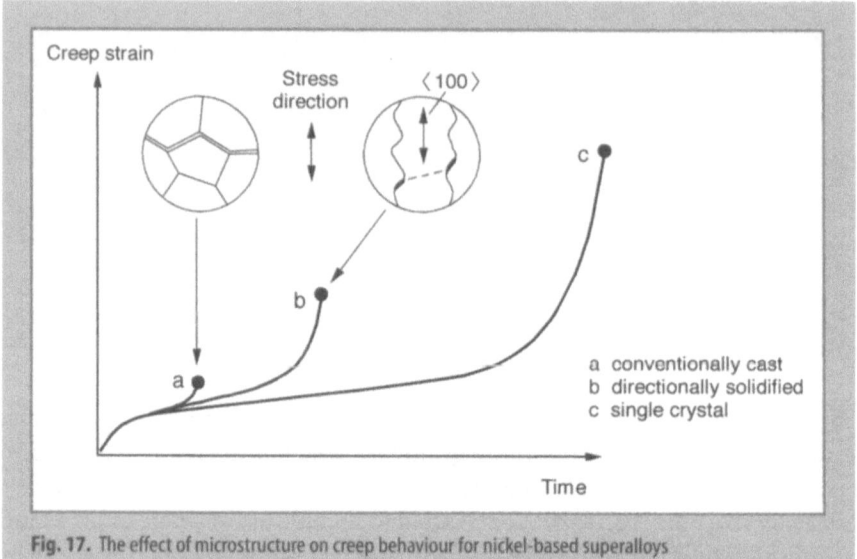

Fig. 17. The effect of microstructure on creep behaviour for nickel-based superalloys

Most high temperature materials rely for oxidation protection on the addition of one or more alloying additions., which oxidise selectively relative to the base metal. The alloying addition must form a stable oxide, which grows at a slow rate (curve a in Fig. 6). Chromium, aluminium and silicon meet these requirements in steels and nickel alloys. However silicon is not used as a major alloying element because the amount which would be necessary to form a continuous external oxide scale would not normally be acceptable from the mechanical behavioural standpoint.

The composition ranges in the Ni-Cr-Al system over which Al_2O_3 and Cr_2O_3 predominate are shown in Fig. 18. Similar data are available for the Fe-Cr-Al system.

In order to avoid internal oxidation and ensure external Cr_2O_3 scale formation, it is necessary to add approximately 25% Cr in Fe and Co and 20% Cr in Ni alloys, although the amount depends on several factors including oxygen partial pressure and temperature. Chromium additions markedly decrease the amount of Al necessary to form external Al_2O_3 scales on Ni, Co and Fe base alloys, as shown in Fig. 18. Al_2O_3 scales grow more slowly than Cr_2O_3 scales – Fig. 19, as predicted by basic diffusion data, and are more protective under oxidising conditions.

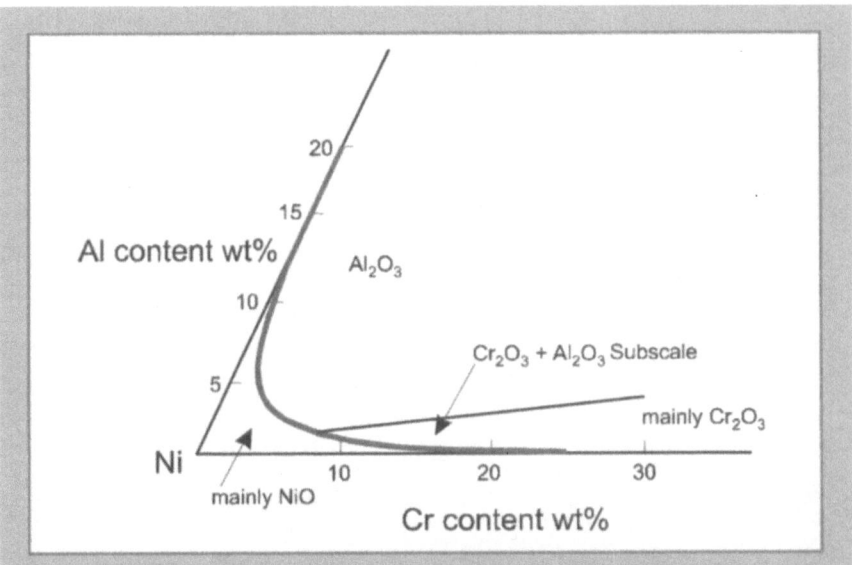

Fig. 18. Simplified phase diagram showing stable fields for Al_2O_3 and Cr_2O_3 at 1100 °C

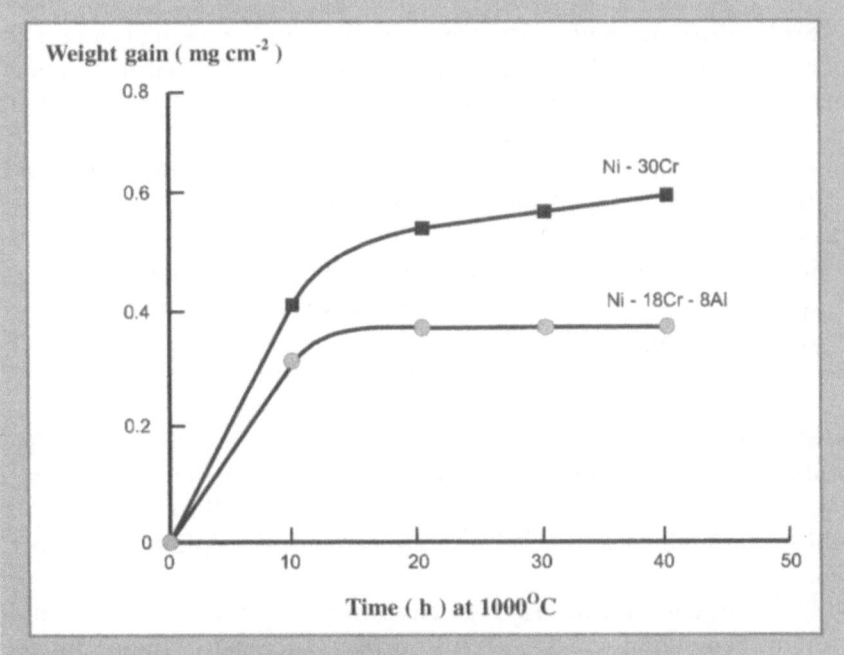

Fig. 19. Comparison of oxidation rates of nickel alloys forming Cr_2O_3 and Al_2O_3 scales

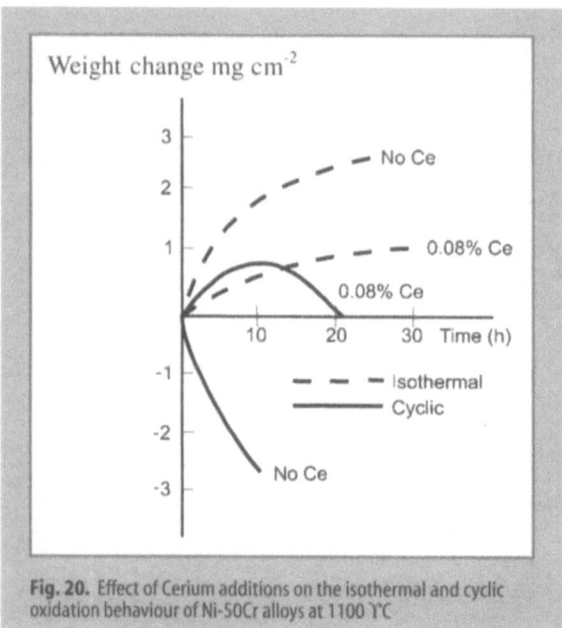

Fig. 20. Effect of Cerium additions on the isothermal and cyclic oxidation behaviour of Ni-50Cr alloys at 1100 ˚C

Scale adherence and resistance to spalling and cracking is a very important factor in oxidation resistance, particularly if thermal cycling is a major operational consideration. Adherence can be significantly improved by small additions of "active" elements such as yttrium, lanthanum and cerium. The effect of cerium additions to Ni-50% Cr alloys is illustrated in Fig. 20 [7]. Mechanisms for these effects have been proposed [8, 9] and include:
- improved chemical bonding across the alloy/oxide interface
- development of "pegs" of reactive metal oxide at the interface
- removal of impurities such as sulphur

Several commercial materials contain such additions.

While Al_2O_3 scales provide better protection against oxidation, Cr_2O_3 scales are better in contaminated environments [10]. The superiority of Cr_2O_3 forming alloys in hot corrosion resistance is illustrated in Fig. 21. Alloys containing less than 15% Cr appear very susceptible to hot corrosion attack.

In sulphidation with H_2/H_2S environments, increasing chromium content generally increases corrosion resistance. Low chromium steels

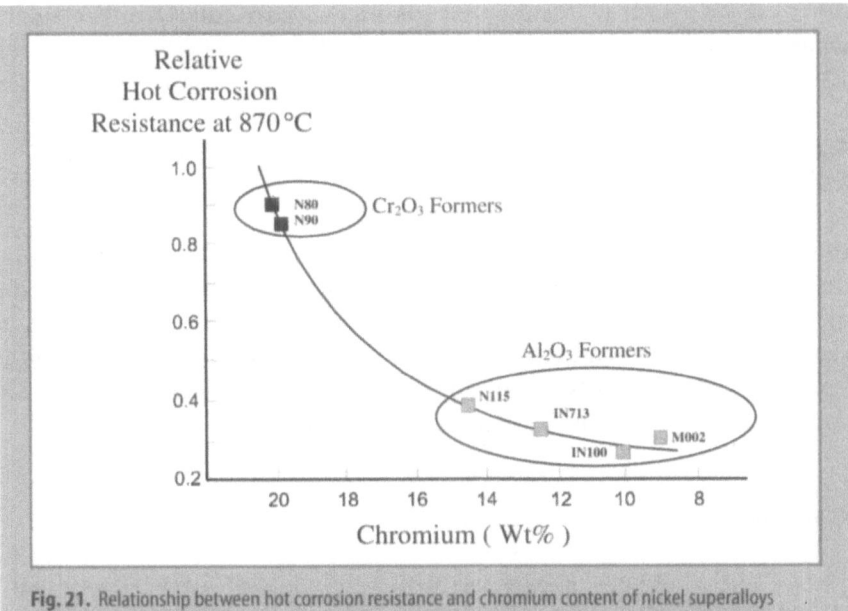

Fig. 21. Relationship between hot corrosion resistance and chromium content of nickel superalloys

have been reported to perform most poorly, 12–16% Cr stainless steels were better while austenitic stainless steels were the most corrosion resistant. Thus in some situations, the solution may be to select the optimum material from the available candidates. This may not provide the required level of environmental protection however and much work has been carried out to develop materials with fundamentally superior environmental resistance. In the petrochemical industry, for example, it has been found that additions of niobium, tungsten, molybdenum and silicon to the standard Fe-25Cr-20Ni alloy used in ethylene pyrolysis furnace tubes resulted in significantly improved carburisation resistance, with silicon being particularly effective [11].

As noted earlier the prime concern with the base material of a component is normally to satisfy the mechanical requirement. The environmental resistance is optimised consistent with this, and may prove to be adequate for the service duty and life. Where this is not the case, surface treatments or coatings are applied. These coatings must be chemically and physically compatible with the base material. They are selected to provide the required environmental resistance, with respect to corrosion, wear, erosion and thermal insulation and the deposition process used must take account of component geometry, particularly in the region where the coating is required. Coatings are discussed further in Chapters 12 and 16.

4.2
Ceramic Materials

In contrast to metals, the major problem with ceramics is a basic lack of dislocation movement resulting in poor defect tolerance. Thus, while microstructures can be controlled to increase temperature capability, increase in defect tolerance is essential for most engineering applications. Ceramics are inherently strong in compression and weak in tension and component design should reflect this, where possible.

4.2.1
Phase Control

The earliest engineering ceramics were multi-phased materials and contained a glassy phase as a consequence of the sintering additions used in manufacture. Alumina, probably the first engineering ceramic, contained 5–15% SiO_2 for this purpose. The strength and temperature capability are dependent on the amount of the glassy phase, as shown in Fig. 22.

Sintering aids are also used in the manufacture of silicon nitride [12]. As is shown in Fig. 23, the creep strength is increased by increasing the viscosity of the glassy phase and by subsequent crystallization of the glass.

4.2.2
Defect Tolerance

The major factor limiting the use of ceramics in stressed applications is poor defect tolerance. Because of a lack of dislocation movement in practical ceramics, they cannot redistribute stress by plastic deformation. Ceramics therefore are much more sensitive than metals to local stress concentrations, which may arise as a consequence of design/operation and to defects which may be introduced during manufacture. Two major approaches have been adopted in the development of engineering ceramics for use in stressed applications at high temperature. The first approach involved accepting the brittleness of ceramics and developing manufacturing and inspection technologies to eliminate unacceptable

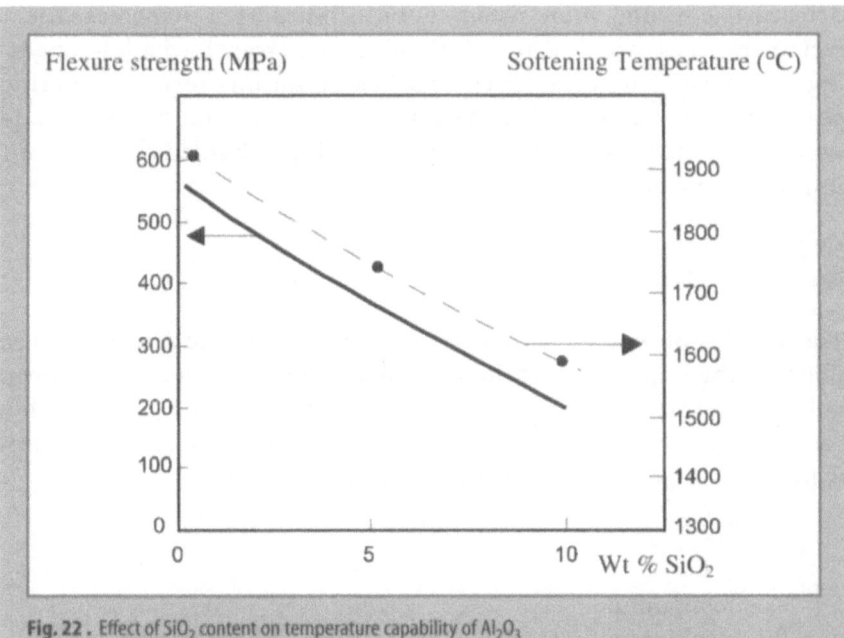

Fig. 22 . Effect of SiO₂ content on temperature capability of Al₂O₃

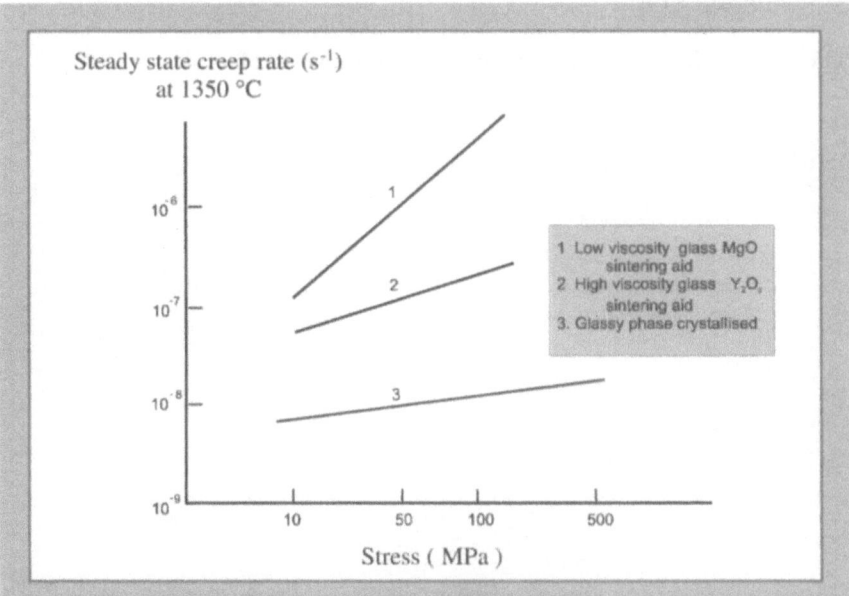

Fig. 23 . Effect of additives on creep properties of silicon nitride based ceramics

defects. The second, more recent, approach has been to increase the defect tolerance of ceramics by techniques such as transformation toughening and developing ceramic composites in various forms. The critical stress intensity K_{1c} of Al_2O_3 can be increased by a factor of 4 by the addition of SiC whiskers and the increased toughness has led to significantly improved tool performance.

4.2.3
Thermal Shock Resistance

The inability of ceramics to redistribute stress by plastic deformation emphasises the importance of their physical properties in applications involving thermal cycling. Hasselman [13] defined the thermal shock resistance of a ceramic in terms of ΔT, the temperature difference which would cause failure under steady heat flow conditions.

$$\Delta T = \frac{\sigma_f (1-v)}{E\alpha}$$

E = elastic modulus
σ_f = fracture stress of material
α = coefficient of thermal expansion
v = Poisson's ratio

Data for major ceramic materials are given in Table 4.

Table 4. Thermal shock resistance of ceramic materials

	Al_2O_3	Si_3N_4 hotpressed	ZrO_2 partially stabilised	Lithium Aluminium Silicate (LAS)
Flexure strength (MPa)	350	700	200	140
$\alpha(°C^{-1})$	7.4×10^{-6}	2.5×10^{-6}	4.5×10^{-6}	-0.3×10^{-6}
E (GPa)	380	310	200	70
$\Delta T(°C)$	95	650	1650	4850

The above data are too simplistic to be used for design purposes but do provide a general comparison of potential candidate materials. The thermal shock resistance of glass-ceramics such as LAS and of fused SiO_2 is well established and is due primarily to low coefficient of expansion. Silicon nitride in the hot pressed condition has better thermal shock resistance than Al_2O_3 because of its lower α and higher strength.

4.3
Composite Materials

Naturally occurring composite materials, such as wood and bone, have been used from prehistoric times. With the aerospace age, the need to "engineer" materials which have markedly superior properties to existing metallic materials has led to the development of composite materials of various types. So far the prime requirement has been to improve properties within the existing temperature capability rather than to increase the temperature capability – Table 5. However, there has been significant recent activity to develop higher temperature capability composites to meet advanced design requirements.

Interactive effects between the constituents of any composite system – fibre, matrix and any coating to control fibre/matrix interface – are of prime importance in determining composite behaviour. Physical compatibility, particularly in coefficient of thermal expansion of fibre and matrix,

Table 5. Major high temperature composite systems

Composite system	Prime requirement	Secondary requirement
Titanium matrix	– Increased strength – Increased stiffness	– Increased temperature capability
Carbon - carbon	– Increased defect tolerance & reliability – Increased stiffness – Increased oxidation resistance	
Ceramic matrix	– Increased defect tolerance & reliability	– Increased temperature capability

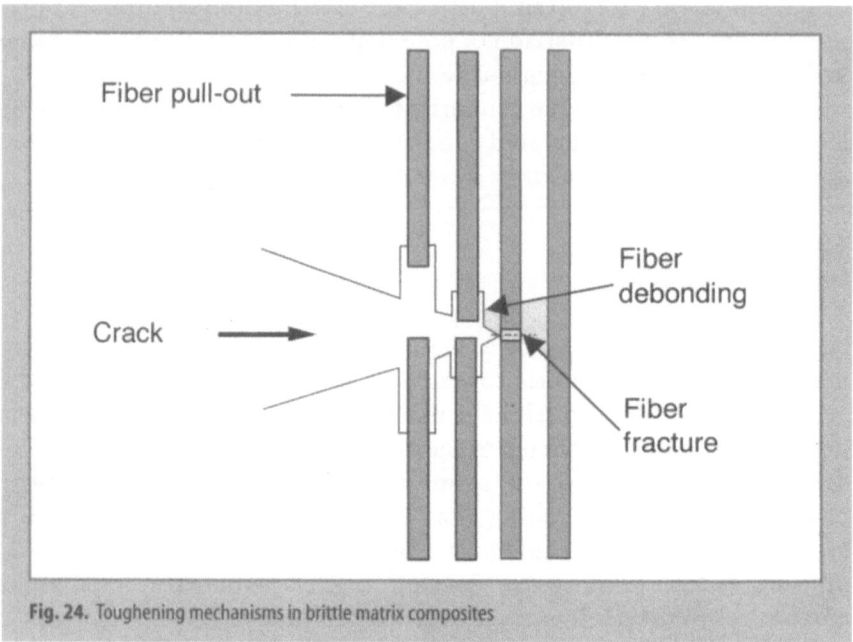

Fig. 24. Toughening mechanisms in brittle matrix composites

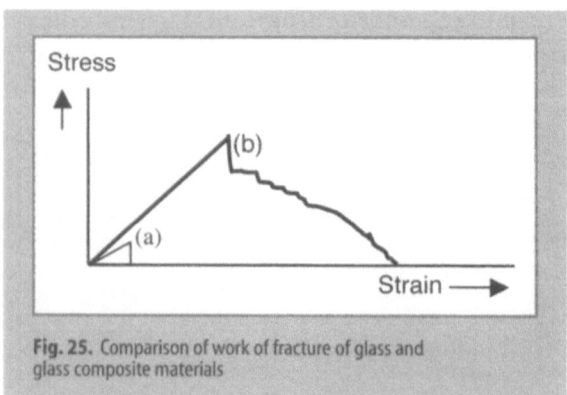

Fig. 25. Comparison of work of fracture of glass and glass composite materials

is essential and chemical compatibility of the constituents is necessary for developing and maintaining appropriate interface properties.

In compliant matrix composites, such as metal matrix composites, a weak matrix is reinforced by strong, stiff fibres. With certain limitations, it is possible to place the fibres precisely where they are needed in a structure. The fibres control the strength and stiffness of the composite. The matrix protects the fibre surface and keeps the fibres in place. It allows stresses to be transferred into and out of the composite and to be distributed among the fibres within the composite. In brittle matrix composites such as ceramic matrix composites, the matrix will crack first, in tension. If the fibre-matrix interface is too strong it will not deflect the cracks and failure will be brittle [14]. With interfaces of appropriate strength, fibre debonding, fibre fracture and fibre pullout all act to absorb energy and reduce crack propagation rate, as illustrated schematically in Fig. 24.

The benefits in terms of work of fracture are illustrated schematically in Fig. 25 for unreinforced glass (a) and a carbon fibre-glass-composite (b). [15]

5 Steels

Steels can be classified into ferritic steels, with a body-centred cubic (bcc) crystal structure and austenitic steels with a face-centred cubic (fcc) crystal structure. Alloying elements can be categorised into those which stabilise ferrite and those which stabilize austenite. Chemical composition determines whether the properties can be modified by heat treatment involving phase change or whether the material consists of stabilised ferrite or austenite. The phase changes include the formation of martensite on cooling from austenite and the subsequent decomposition of the martensite, usually in the case of creep resisting steels, to fine recrystallised ferrite grains with coherent alloy carbides. Steels which can be heat treated in this way are normally referred to as martensitic despite the fact that the final structure is usually not martensite. Austenitic steels are strengthened by the precipitation of carbides or intermetallic compounds.

The relationship between the fundamental types of steel is illustrated by use of a Schaeffler-type diagram, which shows schematically the structure after rapid cooling from elevated temperature – Fig. 26. Elements other than chromium or nickel are included on the appropriate axis as chromium or nickel equivalents depending on their influence on the phases.

General melting and mechanical working aspects of high temperature steels been discussed earlier.

5.1
Ferritic Heat Resistant Materials

Several types of ferritic material have been developed to provide resistance to high temperature corrosion and oxidation. They are characterized by having low carbon and high chromium content [1] and because chromium is a ferrite stabilizing element they cannot be strengthened by

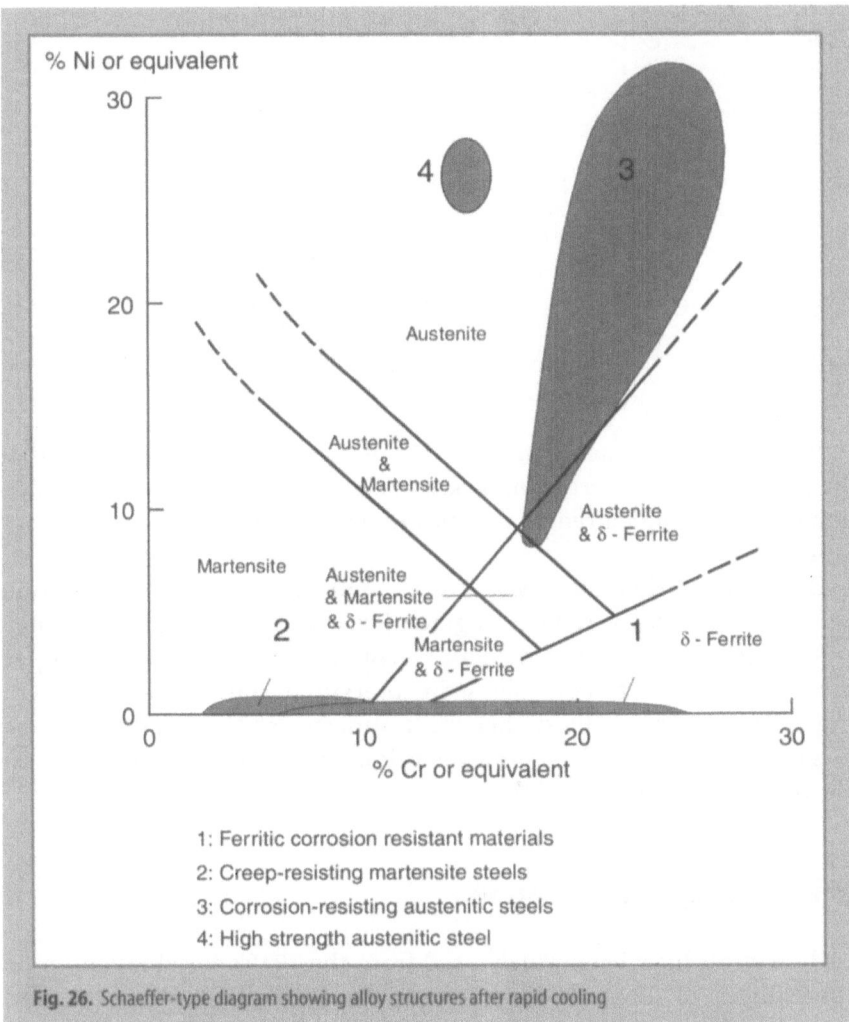

Fig. 26. Schaeffer-type diagram showing alloy structures after rapid cooling

heat treatment. Consequently they are used in applications requiring oxidation resistance but little strength. Temperature capability increases with chromium content – Table 6 and is further increased by additions of silicon, aluminium and small amounts of rare earth elements to reduce oxide spallation, particularly during thermal cycling.

The excellent oxidation resistance of the FeCrAl materials is due to the formation of a protective Al_2O_3 film. They were first used for electric re-

Table 6. The chemical composition and temperature capability of ferritic heat resistant materials

Material	Temperature capability (°C)	Applications
Fe-6Cr	650	Petrochemical plant
Fe-10Cr-2Si	750	Internal combustion engine valves
Fe-15Cr-4.5Al	1050	Chemical plant
Fe-25Cr-5Al-0.3Y	1400	Heat treatment furnace equipment, heating elements

sistance heating elements in the 1930s and are more cost effective than the more expensive corrosion resistant austenitic steels in some applications. Where higher strength is required it may be necessary to use austenitic steels.

During the early 1950's it was realized that ferritic stainless steels could be used in a wider variety of engineering applications if the toughness and ductility, including weld ductility, could be improved. These properties were improved by controlling the interstitial elements carbon and nitrogen, particularly in the higher chromium steels [2]. With the introduction of various new melt refining techniques in the 1960's, this was possible on a commercial scale [3].

5.2
Creep Resisting Martensitic Steels

Carbon steels have been widely used from the 1920's for tubes and pipes operating up to about 400°C in power generation plant. Above that temperature low alloy steels are required. There are several types of commercial low alloy steel. Molybdenum steels containing around 0.5% Mo are used up to about 480°C, with solid solution strengthening being the major factor in developing creep strength [4]. Low creep ductility could be a problem with this type of steel and the addition of 1% chromium was found to give superior creep strength with improved ductility. The addition of vanadium and increase in molybdenum above the level of early low alloy steels gives rise to precipitation hardening, which in steels is usually known as secondary hardening. Strengthening results from pre-

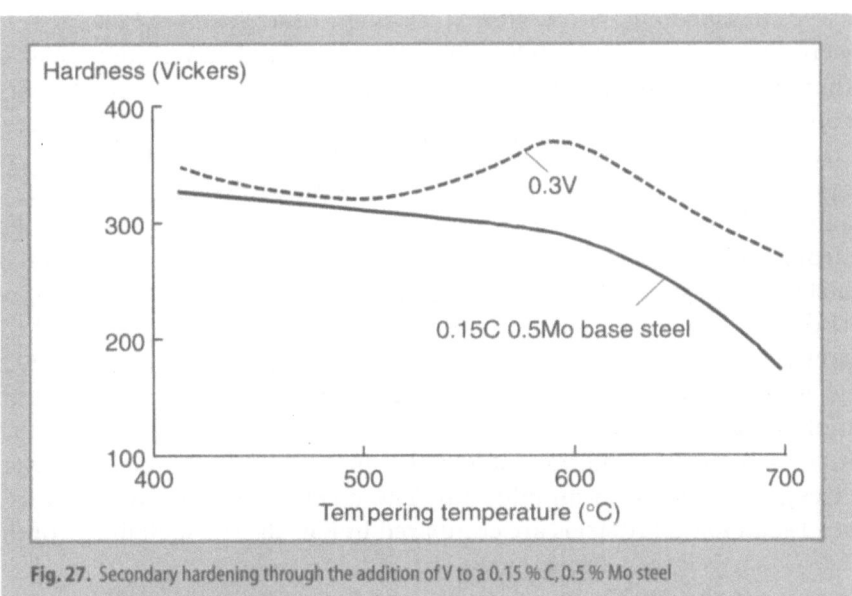

Fig. 27. Secondary hardening through the addition of V to a 0.15 % C, 0.5 % Mo steel

cipitation of V_4C_3 and Mo_2C and the hardening is illustrated in Fig. 27 on a 0.15%C, 0.5%Mo base steel. Steels containing 1%Cr, 0.5%Mo and 0.2%V can be used up to about 580°C and have been used for steam turbine rotors and for turbine discs in early jet engines [5, 6]. Higher creep strength is developed in a steel containing 3%Cr, 0.5%Mo 0.5%W and 0.7%V and this has been used for turbine discs operating at higher stress levels.

Low alloy steels can be subject to embrittlement on long-term exposure in the temperature range 350 – 500°C. This is known as temper embrittlement and arises from the segregation of small quantities of residual impurity elements such as sulphur, phosphorus, tin, arsenic and antimony to grain boundaries. It can reduce toughness at lower temperatures and crack resistance at high temperature [7]. Conventionally melted turbine rotors in NiCrMoV steels are restricted to around 350°C operating temperature for this reason. The problem can be avoided by careful selection of scrap material used in the melting charge and by secondary refining processes such as ladle refining [8].

Inadequate oxidation resistance, rather than creep strength, limited the application of low alloy steels at the higher temperatures of turbine discs in early jet engine designs. Steels containing 12% Cr had been used in steam turbine rotors. They had good oxidation resistance, but lower creep strength than the low alloy steels of the time. Stronger 12% Cr steels were thus needed. The higher proof strength and lower coefficient of thermal expansion of ferritic steels would allow significant weight saving compared with austenitic steels which were then available, if the creep strength of 12% Cr steels could be improved. This need led to the development of steels such as FV 448 and FV 535 [9]. Molybdenum, vanadium and niobium were required for secondary hardening but because they are strong ferrite-forming elements, they had to be balanced by addition of nickel and some reduction in chromium content to avoid the formation of δ ferrite with associated loss of strength. Strengthening involved the precipitation of a M_2X phase at tempering temperatures around 650°C where X is a combination of carbon and nitrogen. Tungsten, vanadium, niobium and nitrogen stabilise M_2X. In FV 535 the strength was further increased by the addition of cobalt and the grain boundaries were strengthened by a boron addition. The creep properties of the 12%Cr steels and low alloy steels are compared in Fig. 28. The metallurgy of the 12%Cr steels is well-documented [10, 11].

Initially the 12%Cr steels met the requirements. Eventually however, the need to improve integrity and cleanness could no longer be met by the existing air melting process and vacuum arc remelting was introduced as

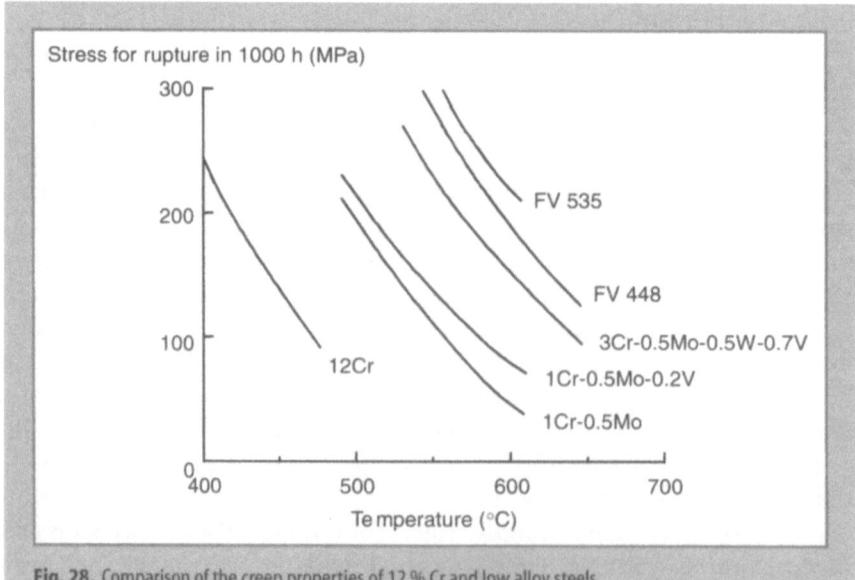

Fig. 28. Comparison of the creep properties of 12 % Cr and low alloy steels

discussed earlier. Electro-slag remelting is also used as an alternative. The need for clean melting techniques was confirmed when an air melted low alloy steel rotor burst in an industrial turbine in the 1950's.

In super-critical steam plant for electric power generation, plant efficiency can be improved by increasing steam temperature and pressure. The limiting factor is the creep strength of low alloy steels in thick section boiler components. First generation boilers used low alloy steels such as 2.5%Cr-Mo but they are limited to subcritical conditions and steam temperatures up to about 560°C. Conditions beyond these values require stronger materials. Early attempts to use austenitic steels resulted in fatigue failures in load cycling operation. Steels such as FV 448 were not suitable for thick section components. The high niobium content in particular resulted in segregation problems in the centre of large rotors. Major reduction in niobium was seen as being necessary, together with a rebalancing of alloying elements to avoid the formation of δ ferrite. Developments based on the 9Cr-1Mo base steel however, which had seen wide usage in the petrochemical industry, led to steels such as P 91 which are now in use world-wide for superheater tubes, headers and piping. P 91 is tough and weldable although heat treatment is required after welding. The creep strength of these steels – (Fig. 29) has allowed the main pressurised parts of super-critical power plants operating at temperatures up to 590°C to be made from martensitic material. As far as 12% Cr steels are concerned, the base 12Cr Mo V steel has been used extensively in German power plants. A Japanese development 12% Cr steel, HCM12A, contains tungsten and copper to provide additional solid solution strengthening. The latest materials may allow operation at steam temperatures up to 620°C with metal temperatures somewhat higher than this. The compositions of the steels referred to are listed in Table 7 and

Table 7. Compositions of 9-12% Cr creep resistant steels

Steel	Fe	C	Cr	Ni	Mo	V	Nb	W	Co	Cu	N	B
FV448	bal	0.15	11	0.7	0.7	0.25	0.3	-	-	-	0.06	-
FV535	bal	0.07	10.5	0.3	0.7	0.25	0.3	-	6	-	0.02	0.005
P91	bal	0.1	9	-	1	0.2	0.06	-	-	-	0.05	-
HCM12A	bal	0.1	12	-	0.5	0.2	0.05	1.8	-	1	0.06	-

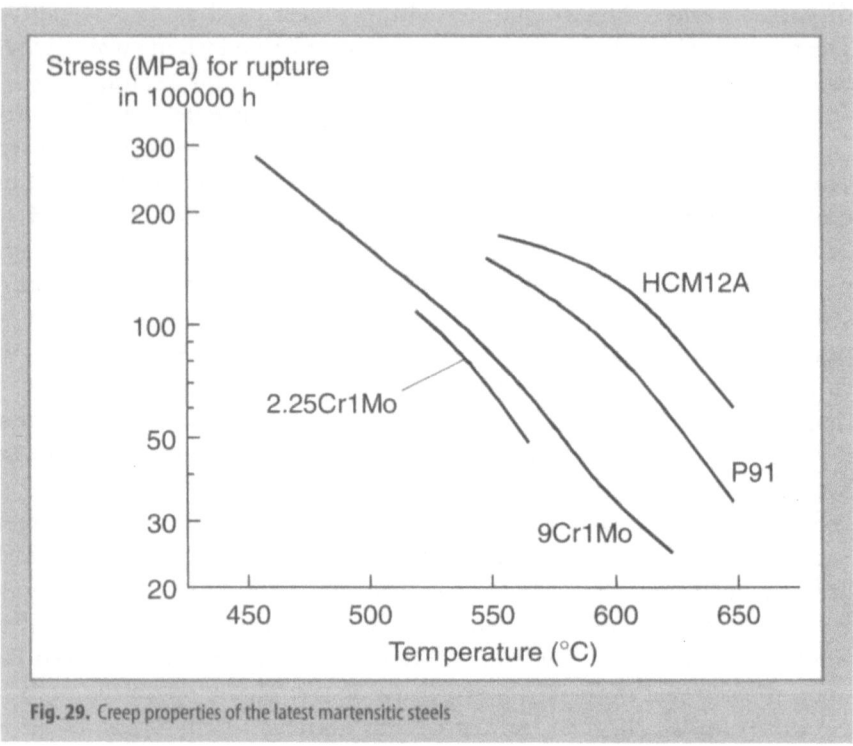

Fig. 29. Creep properties of the latest martensitic steels

their development is documented [12, 13, 14]. Austenitic steels will be required for higher temperature operation.

5.3
Austenitic Steels

Austenitic steels can be categorized into steels used primarily for corrosion resistance and steels where strength is the prime requirement.

5.3.1
Corrosion Resistant Austenitic Steels

The forerunner of the austenitic steels is the 18% chromium 8% nickel (Type 304) steel. The basic 18/8 steel was prone to precipitation of chromium carbides in the grain boundaries when heated in the temperature range 500/850°C. This produces local chromium depletion, which can re-

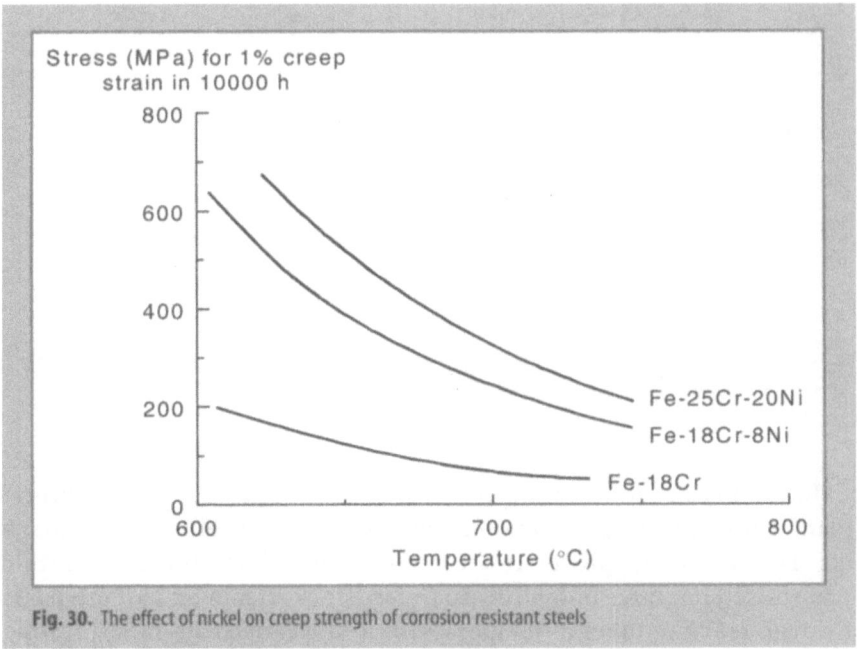

Fig. 30. The effect of nickel on creep strength of corrosion resistant steels

sult in intergranular corrosion if the material is subsequently subjected to corrosive environments. The addition of small quantities of titanium (Type 321) or niobium, in quantities related to the carbon content, results in the formation of TiC or NbC and avoids the problem – the "sensitisation" to local corrosion is reduced. An alternative approach is to reduce the carbon content (Type 304L or 317L).

At given chromium level, austenitic material is stronger than ferritic material – Fig. 30. Increased chromium and nickel contents result in increased strength, with the nickel content having little effect on oxidation resistance – Fig. 31.

The general corrosion resistance of austenitic steels is better than ferritic steels, but the standard austenitic grades have inadequate resistance to stress corrosion cracking (SCC) in aqueous solutions and inadequate resistance to pitting corrosion. Molybdenum is added to improve resistance to pitting (Types 316 and 317) while higher nickel contents as in Incoloy 800 (20Cr/32Ni) gave increased resistance to SCC [15]. Increase in nickel content and molybdenum and copper additions increase corrosion resistance in reducing environments.

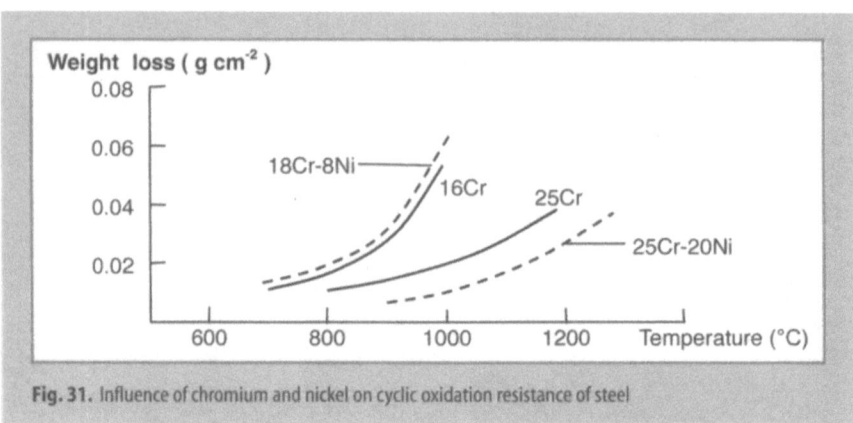

Fig. 31. Influence of chromium and nickel on cyclic oxidation resistance of steel

In the petrochemical industry, continuous high temperature process plant is used for the production of chemicals such as ethylene, methanol, etc. The "cracking" processes are carried out in large furnaces in which pressurised mixtures of hydrocarbon feedstock, steam and air are passed through reaction tubes at temperatures in the range 850–1150°C. Design lives are around 100,000 hours. In pyrolysis (steam cracking), the reaction tubes experience the highest temperatures with the outside of the tube in an oxidizing environment and the inside in a carburising environment. The tube inner surface absorbs carbon from the gas stream itself and the coke layer which gradually deposits. This carbon absorption impairs the mechanical properties and corrosion resistance and thus carburisation is a major life-limiting factor [16]. In reforming (catalytic cracking) the reactions take place in tubes containing the catalyst. The process develops high internal pressures in the tubes. Thus high creep strength to minimise tube wall thickness in order to maximise thermal efficiency is the prime requirement. Carburisation is not a major problem in reforming.

Prior to 1960, tubes were mainly manufactured in wrought stainless steel with limited creep strength. Subsequently, tubes were made by the centrifugal casting process in higher creep strength steels with higher carbon content and additions of niobium, tungsten or titanium. The improvement in creep strength is illustrated in Table 8. It was necessary to compensate for the niobium and tungsten additions by increasing the nickel content to avoid the embrittlement associated with σ phase formation. Sigma phase is a hard brittle intermetallic phase which can form on long time exposure at temperatures in the range 550–900°C.

Table 8. Creep strength of cast austenitic stainless steels

Composition				10^5 hour rupture Life	
Cr	Ni	C	Nb	850ºC	1050ºC
25	20	0.4	-	21 MPa	2.8 MPa
25	35	0.1	1	26 MPa	3.2 MPa
25	35	0.4	1	36 MPa	7.5 Mpa

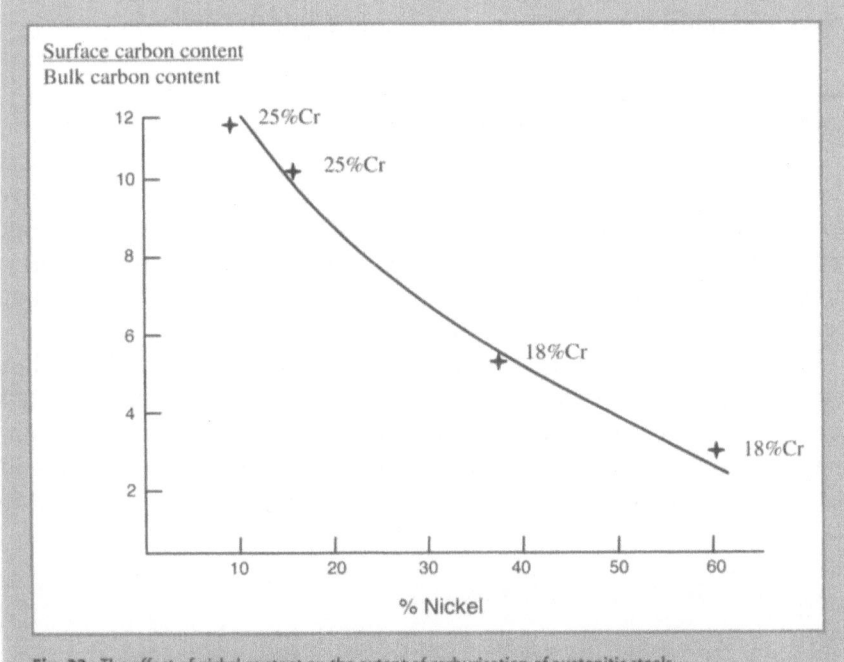

Surface carbon content
Bulk carbon content

Fig. 32. The effect of nickel content on the extent of carburisation of austenitic steels

Increased nickel content is particularly beneficial in reducing carburisation as illustrated in Fig. 32, which shows the effect of alloy composition on the extent of carburisation after 100 hours in a gas-carburizing atmosphere. Silicon additions and carbide stabilizing elements, particularly niobium and tantalum have also been shown to improve carburisation resistance [17].

5.3.2
High Strength Austenitic Steels

The strength of austenitic steels can be increased by precipitation hardening with intermetallic compounds based on Ni_3 (Ti, Al). These materials are derived from Tinidur, a steel developed in Germany during World War II. Good creep strength is possible, A286 for example has a 75°C superiority over FV448 in 1000 hour creep rupture strength. Problems with low ductility were encountered in the early days but were resolved through control of alloy chemistry, melting and forging techniques [18, 19]. Materials such as A286 and Discaloy were used for turbine discs in early US jet engines and are still widely used for discs in industrial turbines. Chemical compositions are given in Table 9.

5.4
Controlled Transformation Stainless Steels

Corrosion resistant steels are necessary in aerospace applications where the temperature capability exceeds that of aluminium and the strength requirement cannot be met by titanium. Examples are jet engine fabrications and the skin of missiles and high-speed aircraft where metal temperatures may reach 500°C and above.

The 12% chromium martensite steels which are available in sheet form have limited formability. Controlled transformation steels have been developed [20] with austenite to martensite transformation temperatures below room temperature. Thus, in the solution treated condition the structure is austenitic and formability and weldability are

Table 9. Compositions of high strength austenitic steels

	C	Cr	Ni	Mo	Ti	Al	B
Tinidur	0.04	15	26	-	2.2	0.15	-
A286	0.05	15	26	1.25	2.15	0.2	0.003
Discaloy	0.04	13.5	26	2.75	1.75	0.1	
AM350	0.10	17	4	2.75	-	-	-
FV520	0.05	16	5.5	1.8	-	-	0.3 Nb
							1.8 Cu

good. Transforming the structure of the formed and welded component to martensite, by either refrigeration to the transformation temperature or by heat treatment to precipitate carbides which raises the
matrix transformation temperature, raises the material strength. The
materials thus combine the formability/weldability of austenitic steels
with the strength of the martensitic steels. The compositions of controlled transformation stainless steels, AM 350 and FV 520 are given
in Table 9.

6 Cast Iron

Cast iron has long been widely used as an engineering material. It was used in high temperature applications in furnaces, boilers, etc. before appropriate steels were developed. It is relatively cheap, is readily castable, easy to machine and its moderate tensile strength combined with high compressive strength make it suitable for applications requiring rigidity and wear resistance. For these reasons it is used in numerous structural and machine applications.

In terms of increasing temperature capability, cast irons can be categorised into grey cast irons, spheroidal graphite irons and austenitic irons.

6.1
Grey Cast Irons

The flake graphite, which counteracts shrinkage during solidification and thus contributes to the castability of grey irons, is detrimental to high temperature behaviour. At temperatures above about 650°C the iron carbide in the pearlite matrix tends to transform to graphite. Oxidation of this graphite and the original flake graphite lead to significant growth [1]. This growth, together with cyclic operational thermal stresses, results in "craze cracking" commonly observed in grey irons operating at high temperature. In CO-containing environments experienced for example in reducing zones close to the grate of a coal fired boiler, carbon deposition can occur within the iron. This can exacerbate the growth and cracking problem during subsequent operation in oxidising conditions. The growth and cracking can be somewhat reduced by small additions of nickel and chromium, together with compensating reductions in silicon. Chromium increases scaling and oxidation resistance while nickel tends to refine the size of graphite flakes. The tensile strength of a grey iron can be significantly increased at temperatures up to 600°C by the addition of 0.75% Ni and 0.45% Cr [2]. Such irons are used for general castings operating at moderate temperature including die casting moulds and automotive brake drums. Stress relief heat treatments may be applied to castings and annealing may be carried out to soften the iron by decomposing massive cementite to improve machinability. Some irons may be rapidly cooled from the austenitic region and tempered to increase the strength [3].

6.2
Spheroidal Graphite Irons

The graphite flakes in grey cast irons contribute to two problems. They contribute to the growth and cracking already discussed. They also act as stress raisers reducing strength and causing brittleness. The addition of magnesium changes the graphite morphology to spheroidal, thus improving the properties over those of flake graphite grey iron, particularly ductility and toughness. The magnesium is normally added as a ladle addition of magnesium or a magnesium alloy. Spheroidal graphite (S.G.) irons are more stable at high temperature because internal oxidation is reduced because of the graphite morphology . Thermal shock resistance is superior to that of grey irons.

S.G. irons have the good casting characteristics of grey cast irons. Nickel additions of 1–2% may be made to increase the tensile properties and heat treatment response. Stress relief, particularly in the case of complex castings, is normally achieved by treatment at 500/550°C. Depending on property requirements castings can be annealed at 850/900°C for maximum ductility, or normalised, or quenched and tempered for maximum strength.

High temperature applications include furnace doors, general furnace castings and refractory tile hangers for furnace walls [4] and a higher silicon iron is widely used for automotive manifolds.

6.3
Austenitic Irons

These irons contain a minimum of 18% nickel. This is sufficient to produce an austenite matrix with either flake graphite or spheroidal graphite. Being austenitic at all temperatures, they do not experience phase

Table 10. Tensile strength of grey and S.G. austenitic irons

Material	Tensile strength MPa		
	20°C	500°C	700°C
Grey iron	250	200	
Grey iron + 0.6 Ni, 0.6 Mo, 0.35 Cr	350	275	
S.G. austenitic iron (20 Ni, 2.25 Cr, 2 Si)	450	350	200

transformations on thermal cycling, and thus have greater stability than other cast irons. The chromium and silicon form protective oxide scales giving oxidation resistance for S.G. austenitics an order of magnitude better than grey irons. The S.G. austenitic irons have better high temperature properties than the flake graphite austenitics. They have higher tensile ductility, 7–25% depending on composition compared with 1–3% for flake graphite austenitics. Nickel is the key alloying element in austenitic irons. The coefficient of thermal expansion is largely dependent on nickel, being at a minimum at 35% nickel. Alloys with 35% nickel have good thermal shock and oxidation resistance.

Flake graphite austenitic irons tend to be used at temperatures up to around 300°C in applications such as pumps, valves and other applications requiring corrosion and erosion resistance. The superior properties of the S.G. austenitic irons allows them to be used in high temperature applications up to 1050°C. The tensile strengths of S.G. austenitic iron and grey iron are compared in Table 10.

S.G. austenitic irons require a casting design approach similar to that used for high strength grey irons, with the casting designed to solidify progressively from light to heavy sections [5]. Stress relieving treatments at 600/650°C are commonly used.

Applications of S.G. austenitic irons include steam turbine parts, steam handling equipment, pump diffusers and impellers, mechanical seals in electric power generation; furnace parts, pump casings and impellers, condenser parts in chemical processing; exhaust manifolds, ex-

Table 11. Compositions of cast irons

	C	Si	Ni	Cr	Fe
Grey Iron	3.0	1.5	-	-	balance
Ni-resist D2	2.8	2.0	20	2.25	balance
Ni-resist D3A	2.5	2.0	30	1.2	balance
Ni-resist D5S	2.0	5.0	35	1.75	balance

haust valve guides, turbocharger housings, nozzle rings and heat shields in internal combustion engines [6, 7, 8].

The major element compositions of typical cast irons are given in Table 11.

7 Nickel Alloys

Nickel-chromium alloys date from the early 1900's when their excellent high temperature oxidation resistance was first recognised. The subsequent development of nickel alloys has been in two main directions. The first has been to improve oxidation and corrosion resistance, with strength being a minor consideration. This was targeted at heat resisting applications in a wide variety of industries. The second has been to increase strength and temperature capability, an objective motivated mainly by the gas turbine.

Melting and mechanical working aspects have been discussed earlier. The alloys which do not contain reactive elements such as aluminium and titanium are air melted and vacuum degassing is commonly applied to improve material quality. Most nickel superalloys must be vacuum melted however. Special casting and working techniques have been developed to improve specific properties of nickel superalloys and these are discussed later.

7.1
Oxidation and Corrosion Resistant Nickel Alloys

The 80 nickel – 20 chromium alloy was the forerunner of the Brightray series of electric resistance heating element materials, the compositions of which are given in Table 12.

The Brightray series of alloys was progressively improved by the additions of silicon and aluminium to increase oxidation resistance and subsequently by the additions of small amounts of cerium and zirconium to improve the adherence of the oxide scale. These improvements increased the maximum operating temperature capability from 950°C for Brightray B to 1250°C for Brightray H. The early alloys tend to be used for domestic heating applications, the later alloys for industrial applications.

Table 12. Chemical composition of oxidation and corrosion resistant nickel alloys

Alloy	Composition, wt%				
	Ni	Cr	Fe	Si	Al
Brightray B	balance	16	24	-	-
Brightray S	balance	20	-	-	-
Brightray C	balance	20	-	1.5	-
Brightray H	balance	20	-	-	3.5
Inconel 600	balance	15	8	-	-
Inconel 601	balance	23	15	-	1.2
Incoloy DS	balance	18	42	2	-
Nimonic 75	balance	20	-	-	0.4 Ti

Concurrently with the Brightray alloys, a series of corrosion resistant alloys was developed for general industrial applications. Inconel 600 is widely used for general industrial furnace parts such as retorts, heat treatment equipment and heat exchangers. Together with Nimonic 75, the forerunner of the Nimonic series of alloys, it was used in gas turbine components fabricated from sheet material such as combustors and exhaust pipes. A more oxidation resistant alloy Inconel 601 has also been used in combustors in solid waste incineration plant, in general furnace equipment and as superheater tube supports in power generation. The benefit of high nickel content in providing resistance to carburisation has been discussed. Alloys such as Incoloy DS are widely used in alternating carburising-oxidising environments in general industry and the petrochemical industry.

7.2
Nickel Superalloys

The so-called nickel superalloys, more than any other material, have made possible the progressive development of the aero gas turbine and the industrial gas turbine. While developed primarily for gas turbines as noted earlier, they are seeing spin-off into other applications where their combined high temperature strength and oxidation resistance are beneficial.

Inconel 600 and Nimonic 75 have been used in exhaust manifolds of aircraft piston engines and more recently in similar applications in high performance sports cars. The precipitation-hardened materials, Nimonic 80 and Nimonic 90 have been used for exhaust valves in piston engines. Nimonic 90 has been used for high temperature springs at temperatures above the capability of spring steels, and various nickel superalloys are widely used in the metal working industry for forging press tools and hot forging dies. Nimonic 90 dies for extruding copper reportedly have three times the life of conventional die steels [1]. Nimonic 80 is used for high temperature bolts in applications such as steam turbine casings where their stress relaxation resistance is superior to steel.

Nickel superalloys are complex materials with alloying additions for various purposes and typically have a combination of high temperature strength and corrosion resistance. The elements which may be present are shown in Fig. 33, which represents part of the periodic table. The heights of the major alloying element blocks give an indication of the amount that may be present and their main functions are indicated. The types of microstructure found in nickel superalloys are illustrated schematically in Fig. 34, with the positions of various alloys indicated. From left to right, there is a progressive increase in γ' forming elements, an associated reduction in chromium and an increase in solid solution strengthening elements.

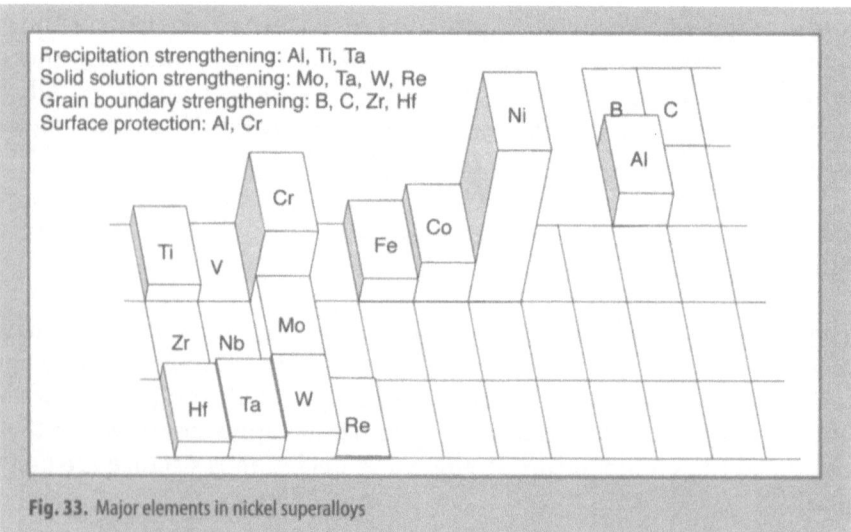

Precipitation strengthening: Al, Ti, Ta
Solid solution strengthening: Mo, Ta, W, Re
Grain boundary strengthening: B, C, Zr, Hf
Surface protection: Al, Cr

Fig. 33. Major elements in nickel superalloys

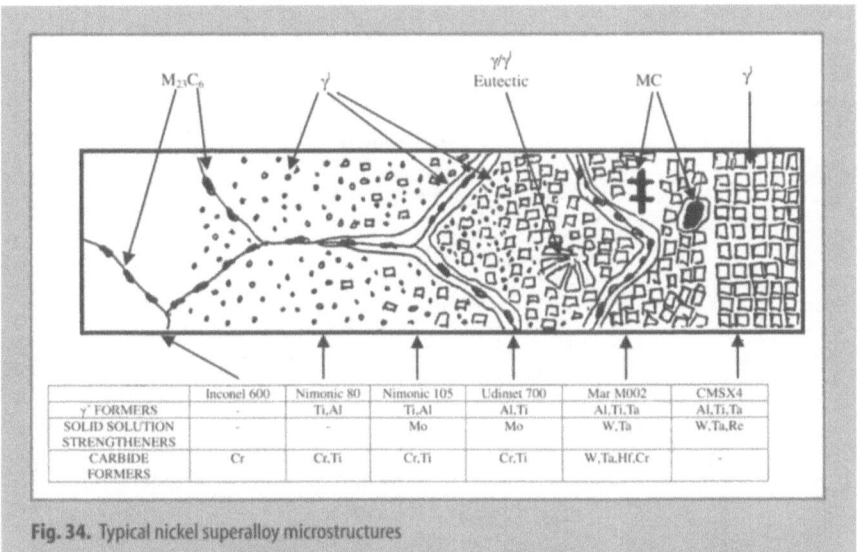

	Inconel 600	Nimonic 80	Nimonic 105	Udimet 700	Mar M002	CMSX4
γ′ FORMERS	-	Ti,Al	Ti,Al	Al,Ti	Al,Ti,Ta	Al,Ti,Ta
SOLID SOLUTION STRENGTHENERS	-	-	Mo	Mo	W,Ta	W,Ta,Re
CARBIDE FORMERS	Cr	Cr,Ti	Cr,Ti	Cr,Ti	W,Ta,Hf,Cr	-

Fig. 34. Typical nickel superalloy microstructures

The development of nickel superalloys and their associated component manufacturing technologies can be considered in three categories [2]. In the early days, alloy development alone was sufficient to satisfy the new design needs. At a later stage in component development, this was no longer possible and the manufacturing process became another key part of the materials technology. When the importance of LCF of disc materials was realised, controlled thermo-mechanical processing became essential for microstructural control for example. Thus in the second category, alloy composition and the manufacturing process are both essential features. The manufacturing process makes the major contribution in the third category, although alloy composition still plays a part in optimising properties and process behaviour. This third category includes directional solidification/single crystal technology and dispersion strengthened materials.

7.2.1
Alloy Composition-Dominated Developments

Turbine Blades
Turbine rotor blades experience high direct stress as a consequence of high rotational speed together with thermal stress arising from the tem-

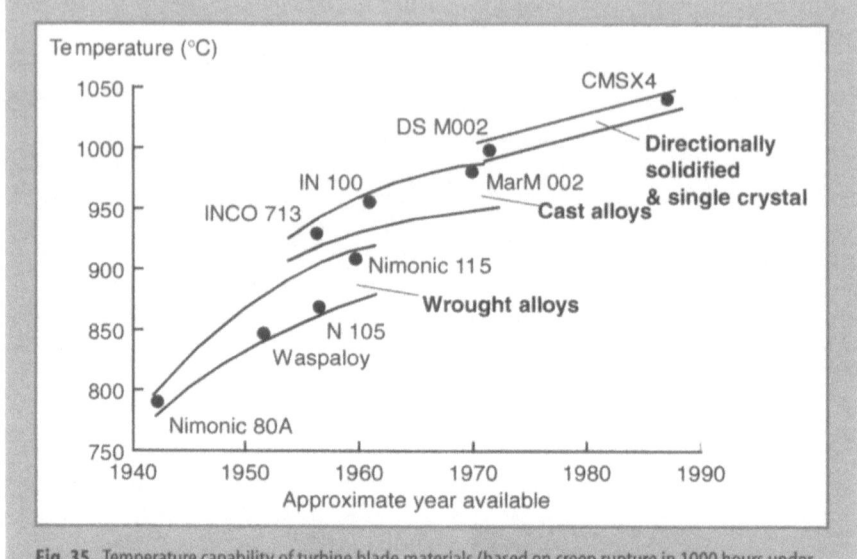

Fig. 35. Temperature capability of turbine blade materials (based on creep rupture in 1000 hours under stress of 150 MPa)

perature transients in the flight cycle of jet engines and load cycling in power generation. Peak metal temperatures exceed 1000°C depending on stage of blade. Turbine stator blades operate in a similar environment but with no direct stress.

Turbine rotor blades have been developed in a process-controlled sequence – Fig. 35 with increased temperature capability in each manufacturing process being dependent on alloy composition.

In wrought alloys [3], the volume fraction of γ' precipitate was progressively increased, cobalt was added to modify the γ' solvus temperature and molybdenum was added for solid solution strengthening. Forgeability became difficult for alloys stronger than Nimonic 115 and Udimet 700. For alloys with higher levels of γ' precipitate an alternative manufacturing technique was necessary - investment casting [4], which had its origins in the lost wax casting process, conceived 2000 years ago. Investment casting was used to produce the supercharger buckets for piston engined aircraft in the 1940's in the cobalt alloy HS 21. This alloy did not contain reactive elements and casting in air was satisfactory. The high aluminium and titanium contents of nickel superalloys required vacuum casting, which was developed in the late 1950's and gave rise to a family of cast superalloys.

Initially the strength was increased by adding more aluminum and titanium to the point where alloys such as IN100 had micro-structures with 60–70 vol% of γ'. Early compositions of IN100 were metallurgically unstable and embrittled because of precipitation of σ phase on long term exposure in the temperature range 650–900°C [5]. It was realised that additional controls on chemical composition were required. From the 1960's internal air cooling was essential in order to achieve the required turbine blade capability and it was necessary to be able to produce complex thin-walled blade castings. Reduced titanium was found to be beneficial for good castability. Later cast alloys such as Mar MOO2 contain tungsten for additional solid solution strengthening together with hafnium which strengthened grain boundaries and improved the consistency of creep behaviour in the intermediate temperature range – a problem which had been experienced by some of the stronger cast alloys [6, 7].

The increase in strength achieved over the years has been accompanied by a progressive decrease in chromium content and consequent decrease in corrosion resistance as noted earlier. The longer life requirement of industrial gas turbine blades required a somewhat higher level of chromium than was necessary in aero-turbine blades. It was found that acceptable creep strength could be achieved with chromium contents of around 15%, with aluminium/titanium ratio ≤ 1 compared with the usual ratio > 1 and refractory metals rebalanced to include molybdenum, tungsten and tantalum. This led to the development of a group of alloys including IN738 and IN792, which have been widely used in industrial turbines [8]. The improvement in temperature capability/corrosion resistance combination is illustrated in Fig. 36. As operating temperatures in industrial gas turbines have continued to increase however, alloys developed for aero-gas turbines are having to be used and this results in a greater dependence on protective coatings.

Shrinkage microporosity in cast components is minimized by feeding system design and casting parameter selection. However, its complete elimination is extremely difficult and most castings contain a minimal level which is acceptable to operating conditions. In some situations, it may be cost effective to eliminate all microporosity in order to maximize component life. This is achieved by hot isostatic pressing (HIP). High temperatures are required and the HIP process may be followed by post-HIP recovery heat treatment, depending on the alloy.

Nickel superalloys contain deliberate additions of elements such as B and Zr which are beneficial in "trace" quantities as noted in chapter 4.

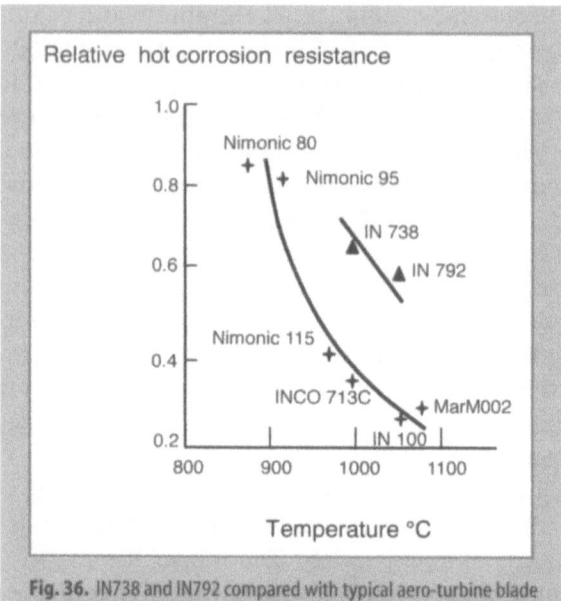

Fig. 36. IN738 and IN792 compared with typical aero-turbine blade alloys (temperature capability based on creep rupture in 100 h under stress of 110 MPa)

Other elements which are harmful when present in "trace" quantities may be present in the initial melt charge material or may otherwise be picked up during manufacture. "Trace" quantities are normally considered to be up to about 500 ppm. The effects of harmful elements have been characterized [9, 10] and steps are taken in manufacture to avoid unacceptable concentrations.

Turbine Discs
Discs operate at high rotational speeds with the highest stresses being in the disc bore where the temperature is lowest. The rim operates at the highest temperatures but with relatively lower stresses. Peak temperatures are in the region of 700°C. The standard manufacturing process was the cast ingot/ forged billet/ forged disc route [11]. Careful control of melting, with vacuum induction melting followed by vacuum arc or electroslag remelting is essential to minimise chemical segregation and ensure forgeability. Alloys such as Waspaloy and Astroloy were developed initially for turbine blade applications and were used subsequently with modified heat treatments for discs. Nickel-iron based alloys Inco 901 and

Inco 718 can be considered as being developed from the precipitation hardened austenitic steel A 286 which has been widely used for discs in industrial gas turbines.

Combustors

Combustors are fabricated from sheet materials and consequently good formability and weldability characteristics are essential. The strength requirement is not high, but, since peak metal temperatures can exceed 1000°C, good oxidation and thermal fatigue resistance are important. Early combustors were fabricated in simple solid solution strengthened alloys such as Inconel 600, Nimonic 75 and Hastelloy X. The need for higher strength combined with weldability motivated the development of the precipitation hardened alloy Nimonic C263, but the strength was achieved at the expense of oxidation resistance. Subsequently alloys such as Nimonic 86, Haynes 230 and the cobalt based alloy Haynes 188 have become widely used. These are somewhat stronger than the earlier solid solution strengthened alloys and contain rare earth element additions to improve oxidation resistance. The benefits of the rare earth elements are probably related to the formation of stable sulphur-containing compounds. It has recently been established that even small sulphur levels reduce the adherance of the normally protective Al_2O_3 scales in high strength nickel superalloys [12, 13] and tests on alloys containing small lanthanum and yttrium additions have shown significantly improved oxidation resistance.

7.2.2
Developments Dependent on Process and Alloy Composition

The recognition in the late 1960's that LCF was a major life limiting feature for turbine discs led to the development of thermo-mechanical processing (TMP) in which forging conditions and heat treatment are controlled to produce a pre-determined microstructure to enhance specific properties [14]. The use of materials behavioural and process models discussed earlier is very important in determining and controlling the TMP process. The design of the forging sequence, including the temperature at which the deformation is carried out in relation to the γ' and carbide solvus temperatures and the deformation rate, is crucial. Control of the forging of ingot to billet and billet to final component are both key factors [15, 16]. Further improvement in chemical homogeneity of the forging stock was found to be necessary for some components and this

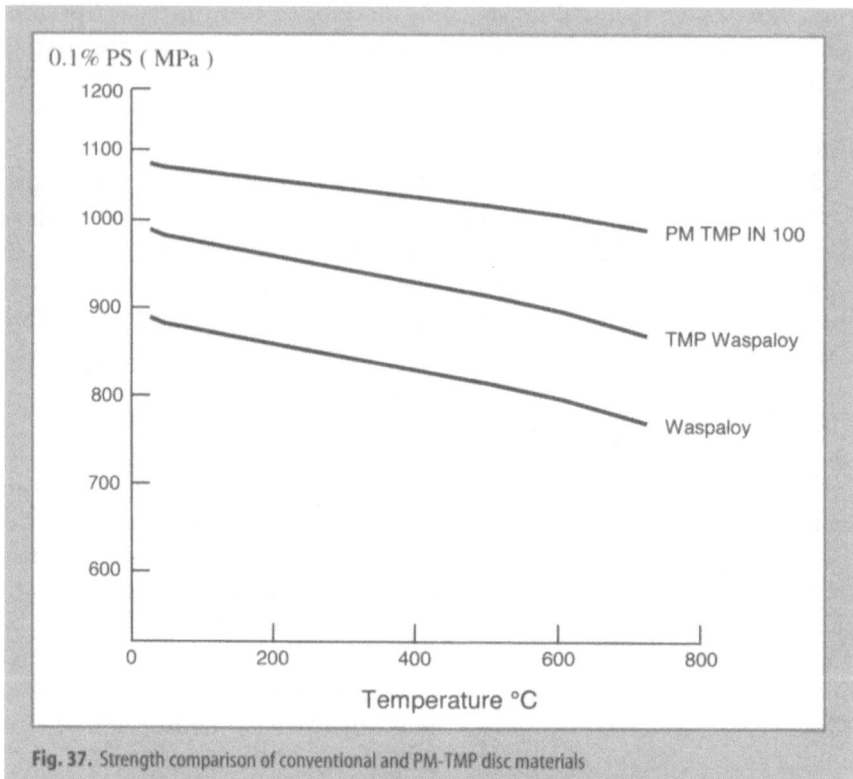

Fig. 37. Strength comparison of conventional and PM-TMP disc materials

led to the introduction of a manufacturing route using powder materials (PM) in the 1970's. Inert gas atomisation is the usual technique for producing pre-alloyed powder [17] and because of the spherical shape and high strength of the alloy, the powder is consolidated into billet by extrusion or hot isostatic pressing [18], before subsequent TMP. Such material has the necessary chemical homogeneity and materials such as IN100 produced by the PM-TMP route had superior tensile properties compared with early alloys – Fig. 37 – and improved LCF properties.

The ultimate fatigue performance of the powder superalloys has been found to be limited by small oxide inclusions inherent in the melting process, which are too small to be detected by current NDE. Thus further progress will depend on the development of cleaner manufacturing processes such as for example, cold hearth refining [19] together with improved NDE capability.

7.2.3
Process Dominated Developments

The ongoing need to increase the temperature capability of nickel superalloys led to the development of directional solidification technology and
dispersion strengthened materials. As was noted earlier, grain boundaries are a source of weakness at high temperature and directional solidification (DS) provides a means of avoiding grain boundaries which are
normal to the major applied stress.

Directional Solidification
Directionally solidified turbine blades contain a series of grains running
from root to tip. The natural crystal growth direction <100> is the direction of lowest Young's Modulus (Fig. 38) and so the thermal stresses generated in service in a DS blade are significantly reduced compared with
investment cast blades. This thermal stress reduction is probably the major benefit arising from DS, although DS material has a higher creep
strength than investment cast material (Fig. 17)

To produce the directional grain structure, a macroscopically planar
liquid-solid interface is maintained in the mould, perpendicular to the
solidification direction, by removing heat from the molten metal in the

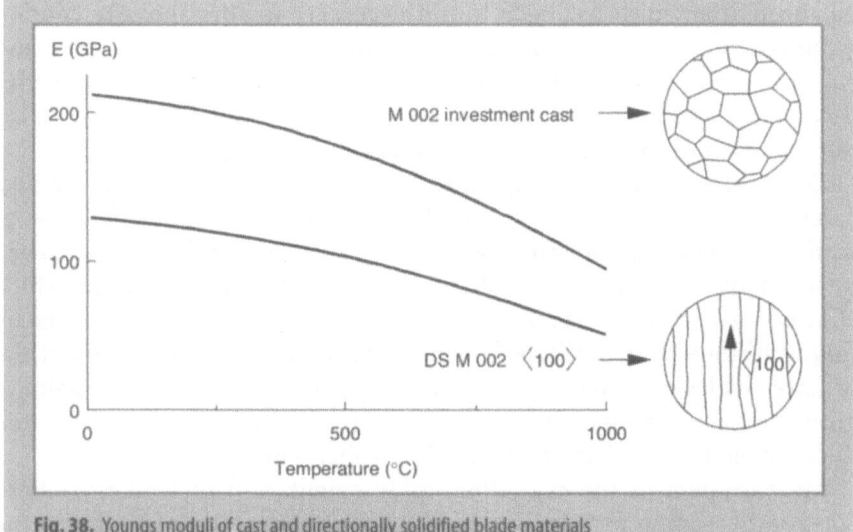

Fig. 38. Youngs moduli of cast and directionally solidified blade materials

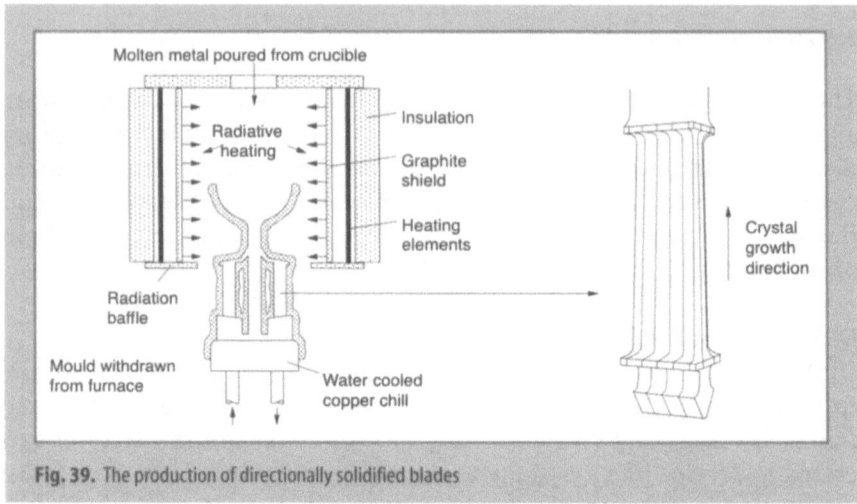

Fig. 39. The production of directionally solidified blades

direction of the desired growth. This is done by heating the mould in a furnace to a temperature above the liquidus and withdrawing it from the furnace at a controlled rate, as illustrated in Fig. 39 [20, 21].

DS blades have been in service since around 1970, largely in alloys which previously were developed for investment castings. Where DS blades have replaced conventionally cast blades, major life improvement has been attained. It has been found necessary for the alloys to contain hafnium to prevent cracking of the longitudinal grain boundaries, and zirconium and silicon have to be limited for the same reason. DS blades up to 625-mm size, root to tip, are being produced for application in industrial gas turbines used in combined cycle power generation.

DS technology has subsequently been modified to produce castings consisting only of one crystal – Single Crystal (SC) – with the objective of further increase in temperature capability [22]. This was initially achieved by the incorporation in the mould of a spiral crystal selection device which allowed only one crystal to grow into the blade cavity. Blades produced by this technique have the <100> orientation along the blade axis but the transverse blade orientation is uncontrolled. It is now standard practice to grow the casting from a seed single crystal positioned in the mould so that the total blade crystal orientation is controlled –Fig. 40.

Measurement of the crystallographic orientation of seed crystals to control the casting process is essential and it is necessary to confirm the

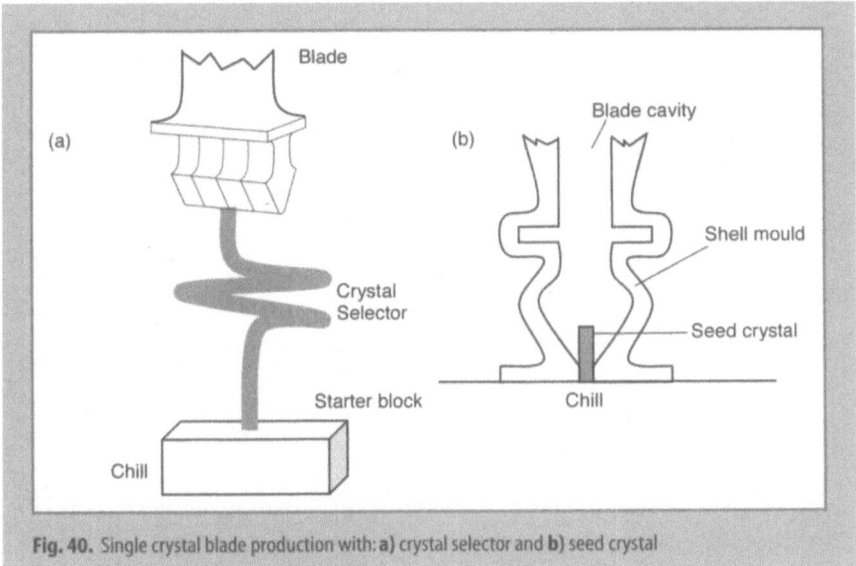

Fig. 40. Single crystal blade production with: a) crystal selector and b) seed crystal

orientation of the blades. The standard laboratory technique for this is Laué X-ray diffraction but the process is slow and not suited to a production situation. Computerized real-time detection techniques have been developed to meet production needs [22].

Since perfect single crystal castings contain no grain boundaries it is possible to remove elements previously added for grain boundary strengthening, such as carbon, boron and zirconium. Their removal raises the incipient melting temperature of the alloy and allows the use of higher temperature solution treatment, thus resulting in increase in creep strength. The latest SC alloys such as CMSX4 and CMSX10 also contain additions of rhenium, which is a very effective solid solution strengthening element, mostly through the formation of small Re clusters [23, 24]. Rhenium also retards γ' coarsening. As illustrated in Fig. 35, these alloys have a significant increase in temperature capability over the "first generation" DS alloys such as DSM002.

It is not easy to maintain ideal control of solidification conditions in the production of large SC blades such as those for industrial turbines. The consequence is that low angle boundaries –boundaries with small degrees of mismatch between adjacent dendrites – are sometimes present in the blades [25]. The greater the degree of mismatch the greater is the loss

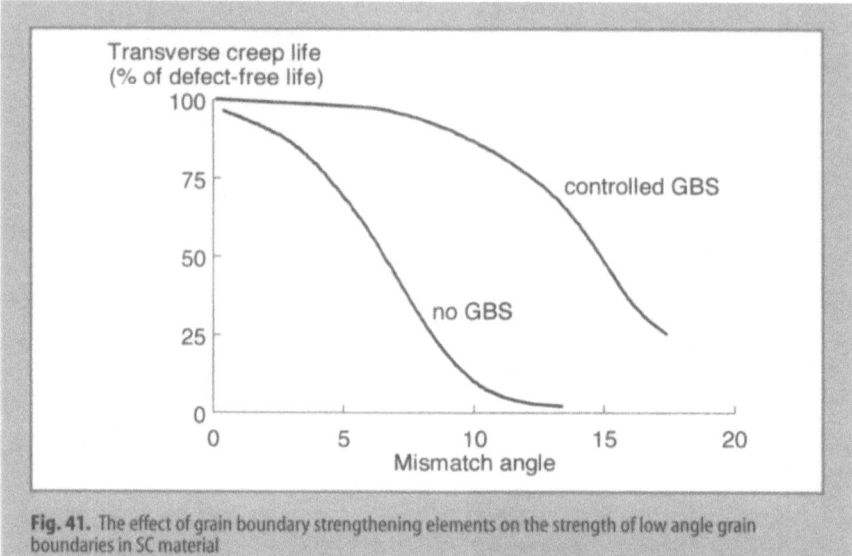

Fig. 41. The effect of grain boundary strengthening elements on the strength of low angle grain boundaries in SC material

in creep strength but this can be offset to some extent by the reintroduction of very small controlled amounts of grain boundary strengtheners (GBS), such as boron and carbon, as illustrated schematically in Fig. 41.

Dispersion Strengthened Superalloys

TD Nickel, the first dispersion strengthened nickel superalloy, contained a 2% thoria dispersion. It was produced by a selective hydrogen reduction process [26]. It was first used for combustion chamber components in the early 1960's. The high thermal conductivity of TD Nickel compared to other sheet superalloys was the major reason for its usage. The poor oxidation resistance was subsequently improved by the addition of 20% chromium in TD Nichrome.

In order to achieve the maximum strength in oxide dispersion strengthened materials, it is necessary to ensure optimum particle size and distribution together with a high aspect ratio grain structure with serrated grain boundaries [17]. Processing is controlled in order to meet these requirements.

A major limitation of the hydrogen reduction process was that it could not produce matrix compositions containing aluminium and titanium. This problem was addressed by the Mechanical Alloying process, which

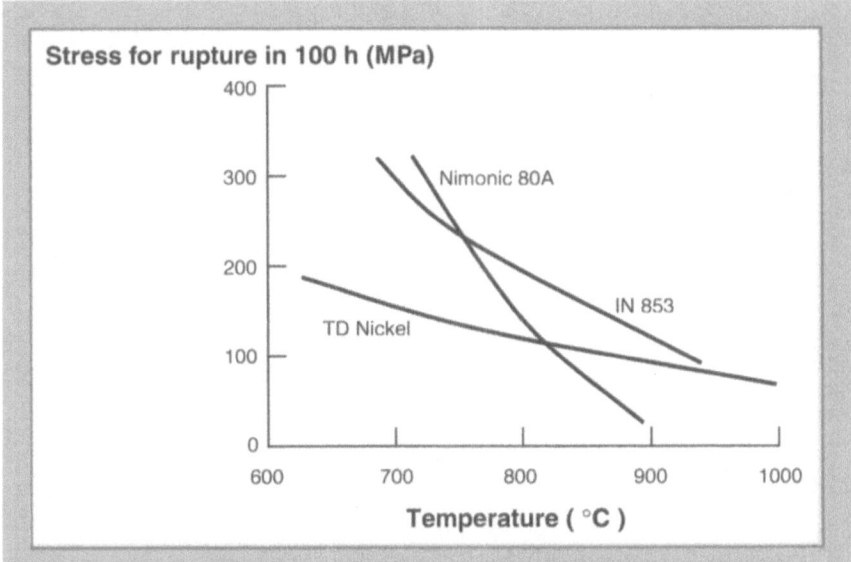

Fig. 42. Effect of combined precipitation hardening and dispersion strengthening on creep strength of nickel superalloy

has the ability to produce nickel alloys with both precipitation and dispersion strengthening [27]. This is illustrated in Fig. 42, which compares the creep strength of the precipitation hardened Nimonic 80 with TD Nickel and IN 853, a nickel base superalloy which is precipitation hardened and also contains a dispersion of Y_2O_3.

The "Mechanical Alloys" offer a unique combination of high temperature strength and oxidation/corrosion resistance. They were developed initially for aero gas turbines and MA754 (Ni-20% Cr containing 0.5% Y_2O_3) has been used in turbine stator blades. Applications of the Mechanical Alloys have since been extended to include industrial gas turbines and various high temperature furnace components. In such cases fusion welding may be an acceptable joining process even though properties are severely degraded locally. Various alternative joining processes have been developed in order to increase joint strength [28].

Materials other than nickel has been dispersion strengthened. In tungsten lamp filaments a 1% ThO_2 dispersion has been used to inhibit recrystallisation and grain growth, in a similar way to the stabilisation of grain size in nickel alloys. A Y_2O_3 dispersion has been used to strengthen the

Table 13. Compositions of various nickel-based superalloys

	Ni	C	Cr	Ti	Al	Co	Fe	Mo	W	Nb	Zr	B	others
Hastelloy X	bal	0.1	22	-	-	-	18	9	0.6	-	-	-	-
Nimonic 86	bal	0.05	25	-	-	-	-	10	-	-	-	-	0.03 Ce
Haynes 230	bal	0.1	22	-	0.3	3	-	2	14	-	-	0.015	0.02 La
Nimonic C263	bal	0.06	20	2.2	0.5	20	-	6	-	-	-	-	-
Nimonic 80	bal	0.07	20	2.4	1.4	-	-	-	-	-	0.07	0.003	-
Nimonic 90	bal	0.07	20	2.4	1.4	16	-	-	-	-	0.07	0.003	-
Nimonic 105	bal	0.12	15	4.7	1.3	20	-	5	-	-	0.1	0.005	-
Nimonic 115	bal	0.15	15	3.8	5	13	-	3.5	-	-	0.05	0.015	-
Udimet 700	bal	0.1	15	3.4	4.3	17	-	5	-	-	0.05	0.015	-
Waspaloy	bal	0.07	20	3	1.4	14	-	4.5	-	-	0.07	0.006	-
INCO 901	bal	0.04	13	3	0.3	-	36	6	-	-	-	0.015	-
INCO718	bal	0.04	19	1	0.6	-	20	3	-	5.2	-	0.003	-
INCO 713C	bal	0.1	12	0.8	6.1	-	-	4	-	2	0.1	0.012	-
IN 100	bal	0.18	10	4.7	5.5	15	-	3	-	-	0.06	0.014	IV
MarM 002	bal	0.15	9	1.5	5.5	10	-	-	10	-	0.05	0.015	2.5 Ta, 1.5 Hf
IN 738	bal	0.1	16	3.5	3.5	9	-	1.7	2.5	0.7	0.08	0.01	1.6 Ta
IN 792	bal	0.15	16	4.2	3.2	9	-	2	3.9	-	0.1	0.01	3.9 Ta
CMSX 4	bal	-	6.5	1	5.5	9	-	-	6	-	-	-	6.5 Ta, 3 Re
IN 853	bal	0.05	20	2.5	1.5	-	-	-	-	-	-	-	1.3 Y_2O_3
MA 956	-	-	20	-	5	-	bal	-	-	-	-	-	0.05 Y_2O_3

iron-base alloy MA 956. This alloy contains 20% Cr and 5% Al and forms Al_2O_3 scales with good oxidation resistance. It has been used in sheet form in industrial burners operating at very high temperatures. Application of the precipitation hardened, dispersion strengthened nickel base alloys has been restricted by the capabilities of the competing DS and SC materials which were developed concurrently.

The compositions of relevant nickel superalloys are given in Table 13.

8 Cobalt Alloys

Cobalt alloys have long been used in high temperature applications – as noted earlier, the alloy HS21 was used for cast supercharger buckets in aero engines in the 1940s'.

Cobalt alloys are solid solution strengthened by the incorporation of the refractory metals molybdenum, tungsten and tantalum. In the case of the cast alloys, they are also strengthened by carbides [1]. It is the presence of carbides which gives the alloys their good castability. The alloys contain sufficient carbon to develop carbides in the as-cast condition and further carbide dispersions are formed during service exposure at moderate temperatures, rather than by the use of specific heat treatments. The carbide strengthening effect is much smaller than that from the γ' precipitation in nickel alloys. Since there is no γ-γ' equivalent in cobalt alloys, the strength of cobalt alloys is much lower than that of nickel alloys, as illustrated in Fig.43. The compositions of cobalt alloys are given in Table 14.

Environmental protection is provided by a high chromium content, in the range 20–30%. This gives rise to a Cr_2O_3 scale which provides good oxidation resistance, although cobalt alloys are not as good as nickel superalloys which form an Al_2O_3 scale. Cobalt alloys are generally more resistant than nickel alloys to the higher temperature type I form of hot corrosion, but appear to be less resistant to the lower temperature type II hot corrosion [2].

The cast alloys typically contain 0.3/0.5% C. HS21 can be regarded as the fore-runner of the high temperature cobalt alloys. The strength and temperature capability have been increased by replacing the molybdenum in HS21 with tungsten and increasing the nickel content to improve ductility in HS31. WI52 has higher tungsten and also a niobium addition, while Mar M 509 contains tantalum giving considerable strength advantage over HS31 [3]. The earlier cobalt alloys are air melted and cast but the subsequent tantalum containing alloys are vacuum melted [4].

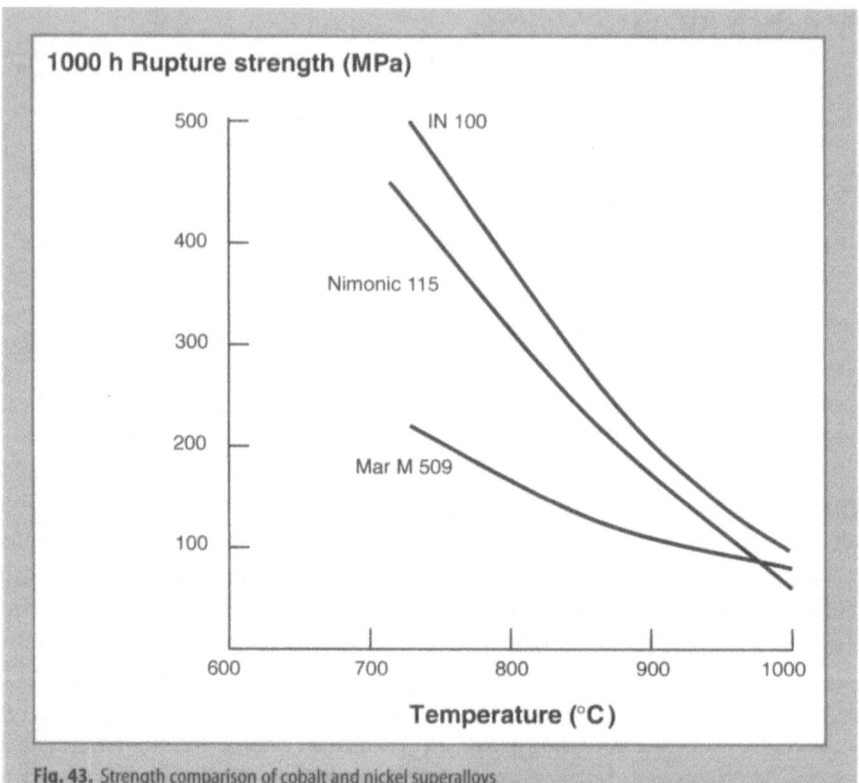

Fig. 43. Strength comparison of cobalt and nickel superalloys

Table 14. Compositions of various Cobalt-based alloys

		Co	C	Cr	Mo	W	Ta	Ni	Nb	others
Cast alloys	HS 21	balance	0.3	28	6	–	–	2	–	–
	HS 31	balance	0.5	25	–	7.5	–	10	–	–
	WI52	balance	0.45	21	–	11	–	–	2	–
	MarM 509	balance	0.5	23	–	7	3.5	10	–	–
Wrought alloys	HS 25	balance	0.1	20	–	15	–	10	–	–
	HS 188	balance	0.1	22	–	14	–	22	–	La 0.05

Because of their property combination, cobalt alloys such as HS 31 and MarM 509 have been widely used in cast turbine stator blades in aero and industrial turbine engines. A major advantage over nickel superalloys in this application is good weldability, which allows weld repair of damaged components at overhaul. Because of their precipitation hardening capability, welding of the stronger nickel superalloys is not easy. Nevertheless, the higher operating temperatures of the latest designs of jet engine are requiring the use of DS nickel superalloys in stator blades, and use of cobalt alloys is thus diminishing in these components.

Cobalt alloys such as HS 25 and HS 188 are used in the form of rolled sheet material. Good weldability and corrosion resistance has led to the use of HS188 in the liners of combustors of aero and industrial turbine engines. A small lanthanum addition increases the oxidation resistance of HS188.

Lower strength alloys have been developed for applications involving severe thermal shock and corrosion, such as grates in oil fired heat treatment furnaces involving repeated heating and quenching. These alloys contain 28% Cr, 20% Fe and 0.1/0.25% C. They have also been used for burners in steel ingot soaking pits heated with a mixture of blast furnace gas and coal gas, for skids for slab reheating furnaces and clinker cooler grates and nose rings for cement kilns [5].

Alloys with higher carbon content and high volume fractions of carbide are widely used as hard facings in high temperature wear protection, either as brazed-in inserts or as coatings. Compositions such as Stellite 6 (Co base -27 Cr -5 W -1 C) and derivatives have been widely used for many years in applications such as mating faces of turbine blade shrouds and exhaust valve seats in diesel engines.

9 Refractory Metals

Refractory metals are normally considered to include the relatively small number of metals which are capable of use at very high temperature [1]. There are four refractory metals with commercial applications – molybdenum, tungsten, tantalum and niobium [2]. They all have high melting temperatures and high densities, which increase with melting temperature. Table 15 gives data for the refractory metals, together with iron and nickel for comparison.

Refractory metals are generally produced by powder metallurgy techniques, with vacuum arc and electron beam melting being used in some cases. Conventional working processes are used to produce mill products but the high hot strengths and low ductilities require particular process technology controls.

The major usage of refractory metals is as alloying additions in steels and nickel superalloys, where they make important contributions to strength and temperature capability. However their high melting points and associated good high temperature strength have led to major com-

Table 15. Properties of refractory metals

Metal	Melting temp. (°C)	Density (g/cm^3)
Niobium	2415	8.6
Molybdenum	2610	10.2
Tantalum	2996	16.6
Tungsten	3410	19.3
Iron	1535	7.9
Nickel	1455	8.9

Table 16. Chemical compositions of refractory metal alloys

Alloy	C	Ti	Mo	Zr	Nb	Hf
TZM	0.02	0.5	Bal	0.1	–	–
C 103	0.05	1	–	0.1	Bal	10

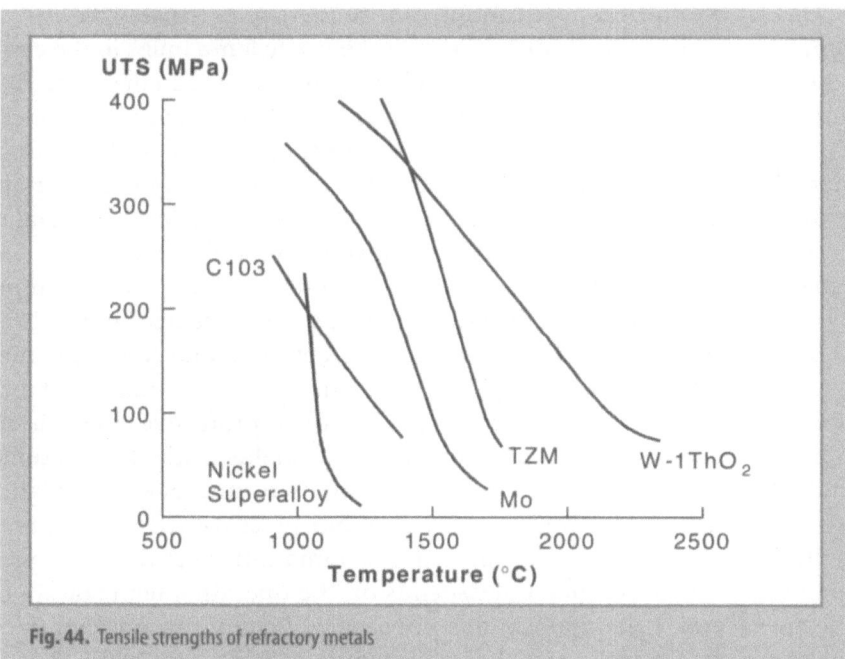

Fig. 44. Tensile strengths of refractory metals

mercial applications of the refractory metals and their alloys. High temperature strength can be increased to some extent by solid solution strengthening and by precipitation hardening. TZM molybdenum alloy is strengthened by small additions of titanium and zirconium in solution and by the formation of small coherent precipitates of TiC and ZrC. Niobium has been solid solution strengthened by additions of hafnium and tungsten and by HfC and ZrC [3, 4]. The niobium alloy C103 is solid solution strengthened by hafnium (Table 16). The strength of refractory metals is illustrated in Fig. 44.

It was noted earlier that oxides such as WO_3 and MoO_3 are volatile and provide no protection against oxidation. Alloys of the refractory metals are also characterised by poor oxidation resistance and this has been a major application-limiting factor. Considerable efforts were made in the 1950's and 1960's to develop oxidation resistant coats. Many of these were silicide based [4] but despite this effort, it is not possible to guarantee long-term protection. Thus, whenever refractory metals are exposed to high temperature, they must be protected against oxidation by either a non-oxidising atmosphere or by vacuum.

One of the earliest applications of a refractory metal was the use of tungsten in the filaments of incandescent electric lamp bulbs in the early 1900's. The filaments reach temperatures up to 2500°C and the protective atmosphere typically consists of 93% argon and 7% nitrogen. In applications where resistance to vibration is a requirement a fine-grained tungsten-1% ThO_2 filament is used with the thoria dispersion inhibiting grain growth. In most applications a filament with coarse axially aligned grains is produced by controlled TMP in tungsten containing potassium and other dopants [5]. In high-pressure sodium lamps, niobium or niobium-1% zirconium tubes are used as current leads and electrode holders [1].

Thermal processing of the newer metals, ceramics and composite materials is carried out at higher temperature than used for traditional materials and increasingly, in vacuum or controlled protective atmosphere. This means that the heating elements, heat shields, racks for handling and positioning the furnace charge and containers must operate at higher temperatures and high purity and low vapour pressure are important requirements. Molybdenum, tungsten and tantalum meet these requirements. The choice of material depends on the operating temperature of the furnace and the applications envisaged. Economic considerations make molybdenum the first choice for temperatures up to 2000°C. Tungsten can be used at temperatures around 3000°C. Tantalum has a temperature capability intermediate between molybdenum and tungsten, but is an excellent getter for oxygen, hydrogen and nitrogen. It can thus greatly improve a vacuum, which can be particularly important when processing titanium, for example. The gettering of the gases gives rise to eventual embrittlement of the tantalum however. Where molybdenum is used, it is normally pure molybdenum but for some applications the higher strength of TZM may be preferred [6].

TZM is used for cores and dies in the die casting of brass, aluminium and steel [7]. Its high temperature strength has more recently led to its use in extrusion and forging dies, such as in the isothermal forging of nickel superalloy gas turbine discs [8]. It is the only available material having the hot strength and creep resistance necessary in this application.

Niobium was considered to have greater potential than molybdenum in aerospace [9] because of its lower density and better oxidation resistance of its alloys, although still not adequate to be used uncoated. Coatings such as a Si-Cr-Ti slurry coating have been developed [10], and alloys such as C 103 have been used for rocket components and for reheat flaps in military jet engines. Use in such applications is being restricted by the latest high temperature composite materials however.

10 Titanium

The titanium industry dates from the late 1940's and has grown very rapidly since that time, largely because of pressure from the aerospace industry. The motivation for the growth was that titanium is light (density approximately midway between aluminium and iron) and has a high strength/weight ratio. It has good corrosion resistance because of a protective surface TiO_2 film.

10.1
Production

Direct reduction of TiO_2 is not practical because of the deleterious effect of small amounts of oxygen in titanium alloys. In the present production process, TiO_2 is chlorinated to $TiCl_4$ which is purified by distillation. The tetrachloride is then reduced by sodium or magnesium. The reaction product is granular, and is compacted into a consumable electrode, together with alloying additions and arc melted in vacuum to produce an ingot [1]. Ingot homogeneity is important because chemical segregation can lead to variation in heat treatment response [2]. Certain alloying elements are particularly prone to segregation and can cause defects which result in reduction in LCF life [3]. Exogenous defects such as hard alpha areas and high density inclusions may arise from contamination of scrap material used in the initial consumable electrode. They can lead to premature component failure unless avoided by manufacturing control or eliminated by inspection.

The initial VAR ingot is re-melted in the consumable arc furnace to improve product quality and recently triple melting has been introduced for the highest duty aerospace components. The ingots are initially worked by press forging with secondary working by extrusion, rolling, closed die forging as required by product. Controlled thermo-mechanical processing (TMP) may be necessary for certain products, such as compressor discs in gas turbines.

10.2
Alloys

The major applications of titanium are in aerospace and general engineering, with the requirements being very different. High specific strength is the major requirement in aerospace applications while corrosion resistance is of prime importance in general engineering. Thus the commercially pure (CP) grades of titanium are of most interest in general engineering applications. Oxygen is the strengthening element in CP titanium. An addition of 0.1% oxygen in Ti 115 gives an ambient temperature 0.2% PS of 200 MPa and this alloy is used where maximum formability is required. A higher level of oxygen increases the strength but at the expense of ductility – 0.3% oxygen gives a 0.2% PS of 430 MPa. General engineering applications [1] include chemical plant, petrochemical processing and steam condensers and, more recently, heat exchangers in applications involving sea water and polluted estuarine water. Titanium is replacing copper alloys in such situations and there has recently been interest in the potential of Ti 6-4, where the superplastic forming and diffusion bonding capability of the alloy is advantageous.

Pure titanium exists in two forms, a close packed hexagonal α phase up to 882°C and a body centred cubic β phase from 882°C up to its melting point. Alloying elements are of three types. Molybdenum, vanadium and niobium stabilise the β phase. Aluminium stabilises α and tin and zirconium are soluble in both α and β and are broadly neutral in phase stabilisation. Strengthening is largely based on the β to α phase transformation and associated micro structural control [4]. Aluminium, tin and zirconium provide solid solution strengthening. Silicon has a major effect on high temperature properties, both by solid solution strengthening and by precipitation of silicide particles if the solubility limit is exceeded [5].

There are three basic alloy types as illustrated in Fig. 45: α, $\alpha + \beta$ and β. Of these, α and β alloys do not have high temperature capability. Being single phase materials, they cannot be strengthened by heat treatment.

Early high temperature alloys were $\alpha + \beta$ alloys containing both α and β stabilising elements. Ti 6-4 was the most common alloy. It was processed and heat treated in the $\alpha + \beta$ region and the microstructure consisted of α grains with β phase in the grain boundaries. It had a temperature

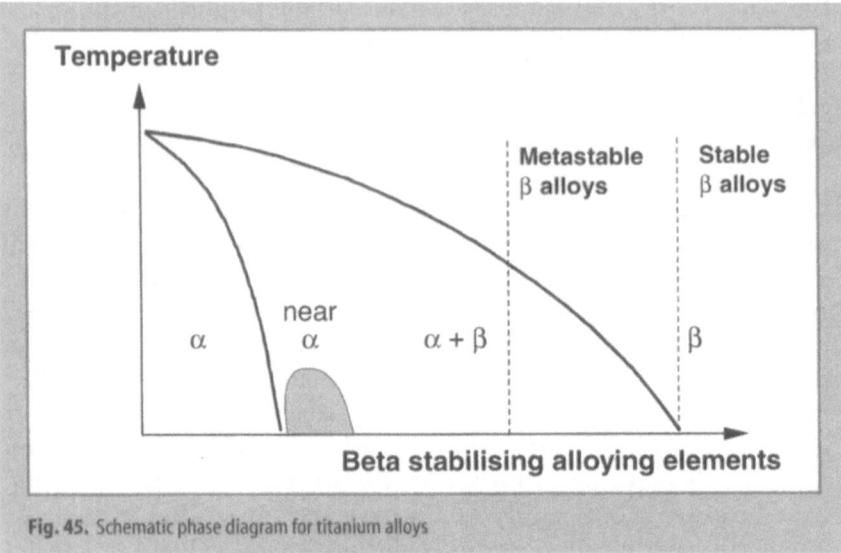

Fig. 45. Schematic phase diagram for titanium alloys

capability of around 350°C and continues to be widely used in lower temperature compressor components. IMI 550 and Ti 6246 are more highly alloyed α + β alloys with higher strength [6].

The need to increase creep strength led to the development of the so-called near-alpha alloys such as Ti 811 and Ti 6242 [7]. They have a compositional balance towards α stabilisation and have a microstructure consisting predominantly of highly alloyed α phase. Further increase in temperature capability was achieved by heat treatment in the β region followed by rapid cooling to produce a martensitic transformed β structure, descriptively referred to as "basket weave". This heat treatment was applied to IMI 685 and the advantage over the α + β heat treatment used for earlier alloys is shown in Fig. 46 [8]. IMI 685 has a temperature capability in excess of 500°C and its use has resulted in major weight saving through replacement of nickel superalloy components in compressors.

While creep and tensile strength were initial targets in alloy development, fatigue strength increased in importance and a balance of properties was seen to be necessary in compressor discs as engine designs progressed. It was recognised that maximising creep properties by changing from an (α + β) micro structure to an acicular transformed β structure could be associated with reduction in fatigue strength. Thus a compromise is required in order to achieve the required balance of properties.

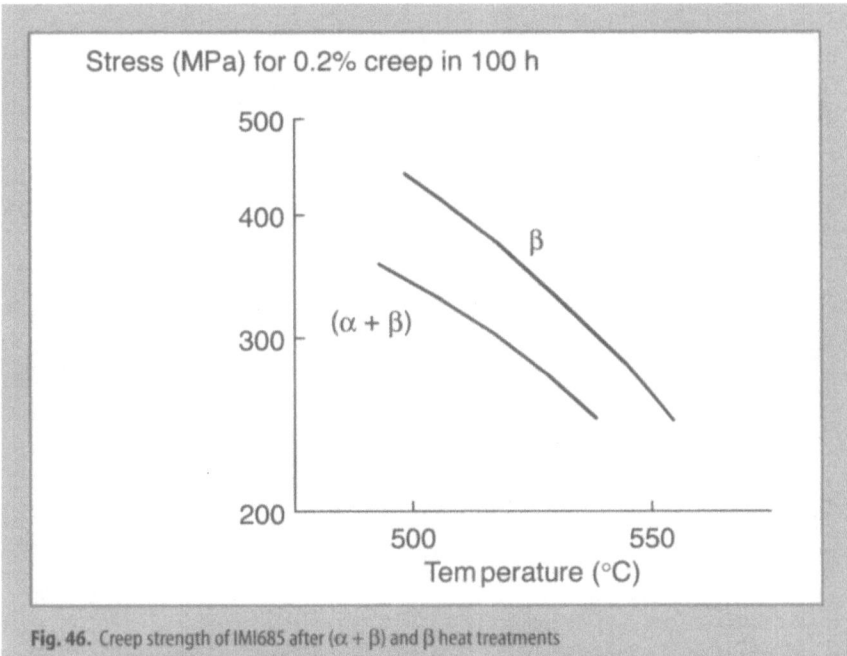

Stress (MPa) for 0.2% creep in 100 h

Fig. 46. Creep strength of IMI685 after (α + β) and β heat treatments

Generally β - heat treated alloys will tend to have higher creep strength and higher fracture toughness. Alloys having (α + β) heat treatment will tend to have higher fatigue strength and tensile ductility. The fatigue strength of Ti 6-4 has been increased by quenching from the (α + β) region to produce a microstructure of α islands in a transformed β matrix [9], and this is now the standard heat treatment for many components in Ti 6-4. Alloy IMI 834, used in compressor discs in the latest jet engines has composition tailored to allow (α + β) heat treatment using a small amount of α phase to pin the grain boundaries. This restricts grain growth and contributes to property optimisation [10].

The increase in creep strength of titanium alloys over the initial CP Titanium alloys is shown in Fig. 47.

There is a further alloy type – metastable β alloys. Relatively high levels of β stabilising elements in α + β alloys can, on rapid cooling, lead to suppression of β to α transformation and retention of metastable β phase to ambient temperature, as indicated in Fig. 45. Alloys with a "molybdenum equivalent" of about 10 are generally considered to be metastable β alloys [11]. Such alloys will precipitate α phase on ageing and significant

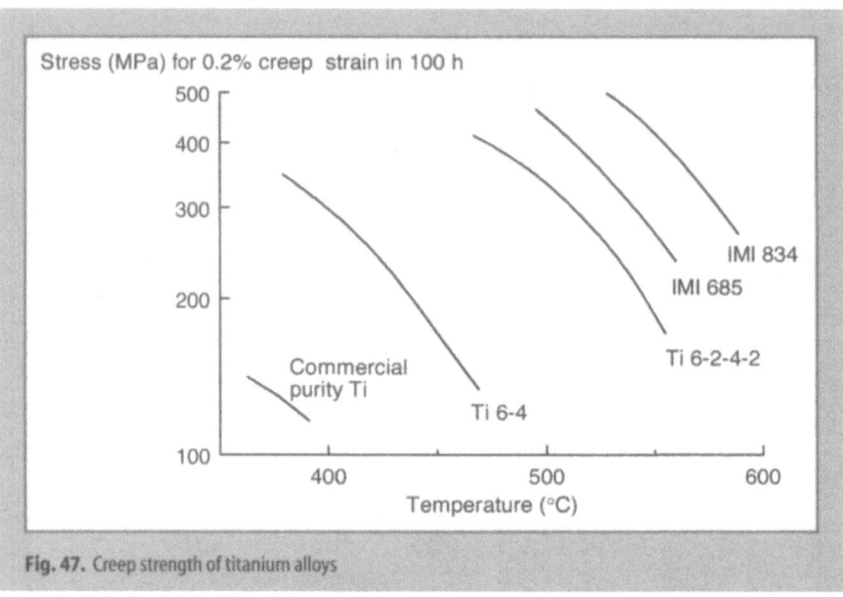

Fig. 47. Creep strength of titanium alloys

strengthening is possible at moderate temperatures. Metastable β alloys have relatively low creep strength and are used in airframe applications. Ti 10-2-3 is used for the landing gear of B777 aircraft with significant weight saving over the original steel [12]. Ti 21S was developed initially to provide oxidation resistant foil for use in Ti MMC material and its good oxidation resistance has led to use in nacelles and exhausts of aero gas turbines.

The compositions of common titanium alloys are given in Table 17.

10.3
Component Manufacture

In addition to the manufacture of components such as gas turbine compressor discs by controlled TMP, [13, 14] two other component manufacturing processes – superplastic forming – diffusion bonding (SPF-DB) and casting – are becoming widely used.

Ti 6-4 exhibits superplastic characteristics because of its duplex (α +β) fine grained microstructure. Sheet material can be formed with low flow stresses at temperatures around 900°C using gas forming methods with modest gas pressures . The superplastic forming can be combined with

Table 17. Compositions of titanium alloys

	Ti	Al	V	Mo	Sn	Zr	Si	Nb	C	Fe
Ti 6-4	Balance	6	4	-	-	-	-	-	-	-
IMI 550	Balance	4	-	4	2	-	0.5	-	-	-
Ti 6-2-4-6	Balance	6	-	6	2	4	-	-	-	-
Ti 10-2-3	Balance	3	10	-	-	-	-	-	-	2
Ti 21S	Balance	3	-	15	-	-	0.2	2.5	-	-
Ti 8-1-1	Balance	8	1	1	-	-	-	-	-	-
Ti 6-2-4-2	Balance	6	-	2	2	4	-	-	-	-
IMI 685	Balance	6	-	0.5	-	5	0.2	-	-	-
IMI 834	Balance	5.8	-	0.5	4	3.5	0.35	0.7	0.06	-

diffusion bonding in the SPF-DB process. Clean surfaces remain oxide-free because oxygen dissolves in titanium at temperatures around 900°C and parent metal strength can be produced in material bonded under pressures of 500–1000 psi [15]. SPF-DB has the potential for significant cost reduction compared with alternative manufacturing processes. The use of SPF-DB to manufacture the heat exchanger duct in Tornado aircraft has been claimed to give a saving of some 36% in cost and weight compared with a conventionally fabricated structure. Fan blades for aero gas turbines are now commonly made by SPF-DB to produce a honeycomb-cored aerofoil structure [16]. Large heat exchangers are currently being manufactured for non-aerospace use in SPF-DB Ti 6-4. Alloys with higher temperature capability than Ti 6-4, such as IMI829 and IMI 834 have been shown to exhibit superplastic behaviour, thus extending the potential for titanium to replace nickel and stainless steel in fabricated structures.

Castings in Ti 6-4 have long been used in relatively low stressed low temperature applications such as aero gas turbine fan frames. The castings have been up to 150kg in weight. Recently several advanced process technologies have, in combination, increased the potential of titanium castings in critical structural applications [17]. These process technologies are vacuum arc and induction skull remelting to minimise contamination in cast titanium, the development of ceramic and metallic mould materials which can contain molten titanium with a minimum interaction and the use of hot isostatic pressing (HIP) to improve casting quality by the elimination of internal microshrinkage and gas defects. Process modelling is a key element in the overall technological capability. Alloys with higher temperature capability than Ti 6-4 are showing promise in castings.

11 Intermetallic Materials

Intermetallic materials are compounds formed by two or more metals in well defined integral proportions and which have a characteristic crystal structure that is ordered up to, or close to, the melting point. Intermetallics generally have high melting points and high resistance to deformation - this leads to the potential for high temperature capability which has motivated much of the recent work.

Serious research on high temperature intermetallics began in the early 1950's and increased significantly around 1970 because of their perceived potential in aerospace. With weight saving being a key requirement, early work concentrated on the aluminide intermetallics based on nickel and titanium. Subsequently intermetallics such as Fe_3Al have been developed because of their potential benefits in replacing steels in various high temperature applications. The high temperature capability of intermetallic materials is unfortunately accompanied by low ductility and toughness at ambient temperatures. Consequently a major objective in the development of practical intermetallic materials has been, and continues to be, to introduce a measure of toughness while retaining the temperature capability. Various approaches are being investigated to achieve the necessary property improvement. These include multi-phase materials concepts, alloying with solid solution elements to modify slip behaviour, strengthening of grain boundaries by micro-alloying, a single crystal approach and the possibility of intermetallic matrix composites.

Some intermetallic materials of interest are listed in Table 18. More recently exploratory work has started on speculative intermetallic systems aimed specifically at very high temperature capability.

Currently certain intermetallics, such as the aluminides, are at the threshold of application. Some of the more speculative intermetallics are at the beginning of their development while there will be others which will not progress beyond the laboratory stage.

Table 18. Properties of common intermetallic materials

Intermetallic	Melting temp. (°C)	Density (g/cm^3)
Ni_3Al	1400	7.5
NiAl	1640	5.86
Ti_3Al	1460	4.2
TiAl	1460	3.9
$MoSi_2$	2020	6.24

11.1
Titanium Aluminides

The major intermetallics of interest in the Ti-Al system are Ti_3Al and TiAl – Fig. 48.

It has been found that in both materials, strength and ductility are enhanced in two-phase structures. In Ti_3Al materials, the two-phased structure has been achieved by adding β-stabilising elements to produce a material bearing a strong resemblance and heat treatment response to conventional $(\alpha + \beta)$ titanium alloys. Microstructural and property modification is therefore possible. The alloy "Super α_2" is the best known Ti_3Al-type alloy, and contains 15%Al, 20%Nb, 3%V and 2%Mo by weight. The alloy has been tested in static structures in jet engines [1, 2]. Its density and creep strength are comparable with the latest titanium alloys. The potential for higher temperature capability is currently limited by a sensitivity to oxygen penetration embrittlement in air at 600/700°C [3].

The γ alloys based on TiAl appear to have significantly higher potential than the α_2 alloys. They have a density of around half that of nickel superalloys with better oxidation resistance than the α_2 Ti_3Al material. Depending on chemistry and microstructure, γ alloys have been classified into three groups – the first generation alloy Ti-48 Al-IV-0.1C, the second generation Ti-48 Al-2Nb-2Cr alloys and third generation alloys. Room temperature elongations up to 4% and fracture toughness values of 10–20 MPam$^{1/2}$ are achievable in the second generation alloys [4]. Processing of the alloys has largely been by conventional methods including casting + HIP and wrought PM. Near-net shape components such as turbine blades, turbochargers and automobile exhaust valves have been made by

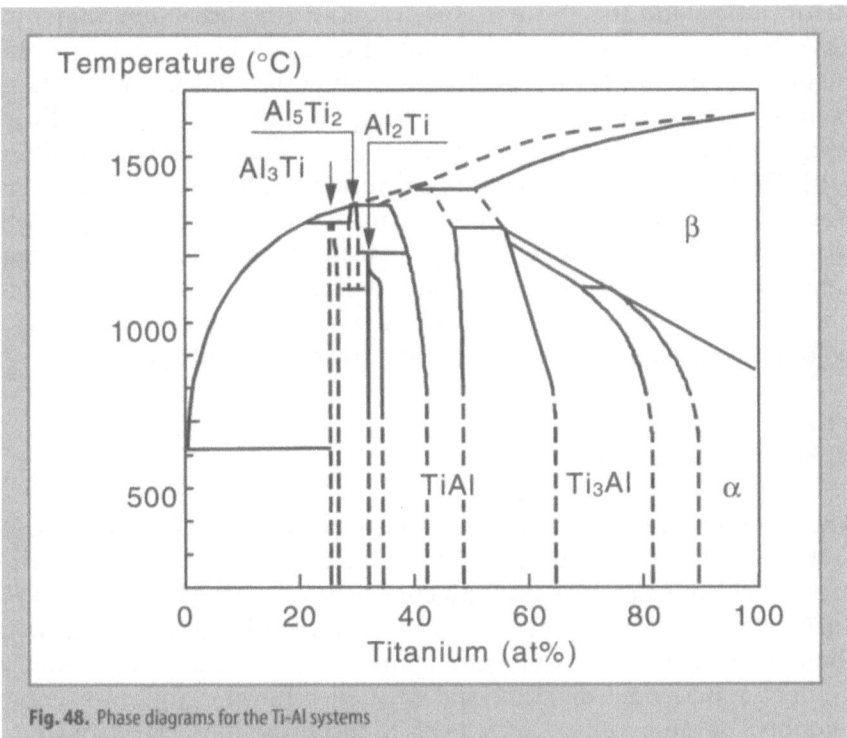

Fig. 48. Phase diagrams for the Ti-Al systems

investment casting. Mechanical working and/or heat treatment are aimed at producing microstructures varying from duplex TiAl/Ti$_3$Al to lamellar TiAl depending on property requirements [5, 6]. However, significant progress is needed before wide commercial application is realised. Uncoated γ aluminides have inadequate oxidation resistance at temperatures above 750/800°C, because they form a mixed Al$_2$O$_3$/TiO$_2$ scale in air [7]. Al$_2$O$_3$ is dominant at temperatures below 750°C, TiO$_2$ is dominant above 800°C. Mixed Al$_2$O$_3$/TiO$_2$ scales are less protective than continuous Al$_2$O$_3$ because TiO$_2$ has a higher rate of growth than Al$_2$O$_3$ and TiO$_2$ can act as a short-circuit transport path for interstitial oxygen and nitrogen dissolution in the alloy with consequent embrittlement. Oxidation resistance can be improved to some extent by alloying, with Nb, Ta, W and Si being beneficial [8]. However, protective coatings are likely to be required and Ti-Al-Cr coatings have shown promise [9].

Despite the ongoing need for further basic work on behavioural understanding, particularly in the area of fundamental mechanisms of deformation and their role in ductility and fracture, some materials have been run in development jet engines. Low-pressure turbine blades cast in a Ti-47Al-2Cr-2Nb at % alloy have been run in a large commercial transport engine and the potential to make significant weight saving relative to nickel superalloy blades has been claimed [10]. Other components such as compressor blades, vanes, casings, nozzle flaps and tiles have been identified as possible applications through rig testing and bench engine testing. Databases and damage tolerances have been assessed. There is significant interest in various automotive engine components, but the over-riding need in this area is for a low cost, high volume manufacturing technology [11].

11.2
Nickel Aluminides

The major intermetallics in the Ni-Al system are Ni_3Al and NiAl – Fig. 49.

The fact that Ni_3Al is the major strengthening phase in nickel superalloys provided the motivation for a considerable research effort from the 1970's [12]. Brittleness has been a problem with polycrystalline Ni_3Al. Although there is an environmental influence, brittleness originates at grain boundaries and single crystal Ni_3Al has shown significantly higher ductility. Boron additions have been found to be most effective in improving tensile ductility and boron also appears to be effective in eliminating environmental embrittlement effects at room temperature [13]. Elevated temperature embrittlement appears to be related to oxygen penetration along grain boundaries and chromium additions reduce the effect. Mechanical properties are influenced by deviation from stoichiometry (Al/Ni ratio) and by alloying additions. Hafnium and zirconium are effective strengtheners. Alloys containing some 16 Al and 8 Cr (atomic percent) with Mo, Zr, Hf and B additions have been developed for structural use at elevated temperature [14]. The alloys with optimum properties contain around 10% of the disordered γ phase. Such materials have seen use in diesel engine components and other applications but general properties are such that they are unlikely to replace nickel superalloys in aerospace applications.

In contrast, NiAl offers a more attractive property combination. It has lower density and better oxidation resistance than Ni_3Al. The major problems with NiAl are brittleness at ambient temperature and low strength at elevated temperature. Significant increase in strength can be

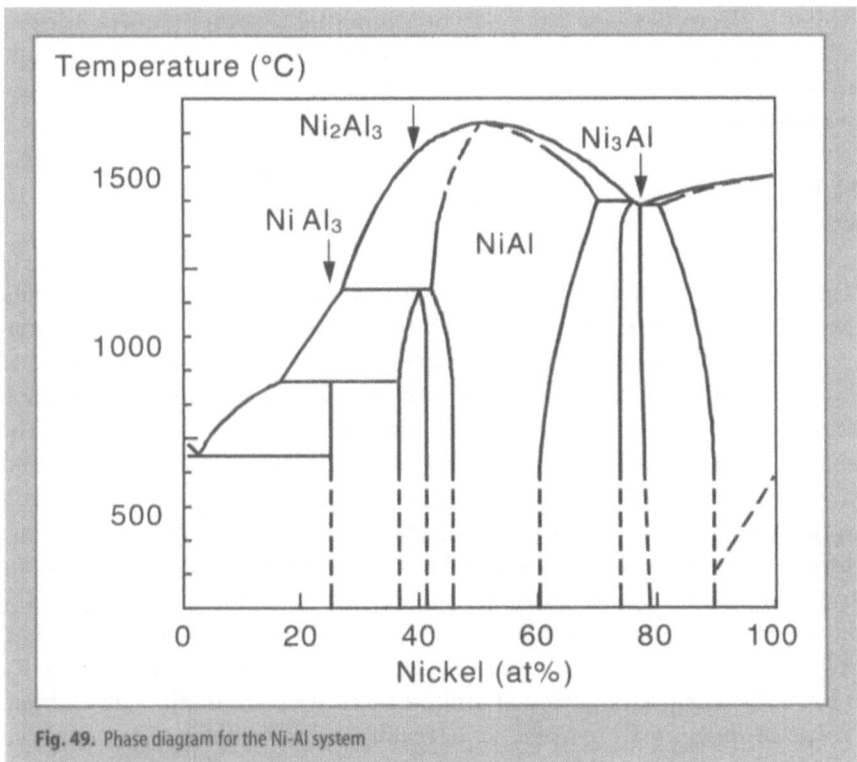

Fig. 49. Phase diagram for the Ni-Al system

Table 19. Creep strength of nickel aluminides (temperature for steady state creep rate of $10^{-7}s^{-1}$ under stress of 100 MPa)

Material	Temperature (^{o}C)
NiAl	600
NiAl/Ni$_2$AlTi	1050
MarM 200	1100

achieved through alloying. Incorporating a second phase, Ni$_2$AlTi, in the NiAl matrix can develop creep strength approaching the nickel superalloy MarM 200 (Table 19) but brittleness remains a problem. Ductility has been developed in single crystals, but not in alloys having high creep strength. Promise exists but much fundamental work remains to be done [15]. Casting of NiAl compositions is difficult, with excessive segregation, but such problems could be overcome by using a powder metallurgy approach [16].

11.3
Iron Aluminides

Iron aluminides have excellent oxidation and corrosion resistance, with lower rates of attack than the best commercial ferritic heat resistant materials. They offer the potential of lower material cost, low density and conservation of strategic elements. Major problems are low ductility and brittleness at ambient temperatures and poor strength above 600°C. The major cause of the brittleness has been identified as environmental effects due to moisture in the air [17]. This understanding has led to ductility improvement by alloy additions to reduce hydrogen solubility and diffusion, improvement in grain boundary cohesion by microalloying and refinement of grain structure by thermo-mechanical treatment. The development of strong ductile Fe_3Al and FeAl based alloys is summarised [18]. A Fe_3Al based alloy containing 28 Al-5 Cr-0.5 Nb-0.5 Mo-0.1 Zr-0.2 B has 13% tensile elongation at ambient temperature in air. The ambient temperature elongation of FeAl can reach similar levels in alloys containing Zr, Mo and B. Ductility can also be improved by grain refinement with TiB_2 additions.

Initial interest in iron aluminides is in aggressive reducing environments such as in coal gasification systems, in which the nickel aluminides do not have sufficient corrosion resistance. A variety of mill products are available, including forgings, extrusions and flat products.

11.4
Speculative Intermetallics

The need for materials for application at very high temperatures has led to research beyond the aluminide systems, into silicides and various other intermetallics including niobium-based trialuminides such as (NbTi-Hf)$_2$Al, beryllides and chromides [19, 20].

Silicon base intermetallics have attracted most interest because of their generally high melting points, good high-temperature oxidation resistance and relatively low density. Probably the most studied compound has

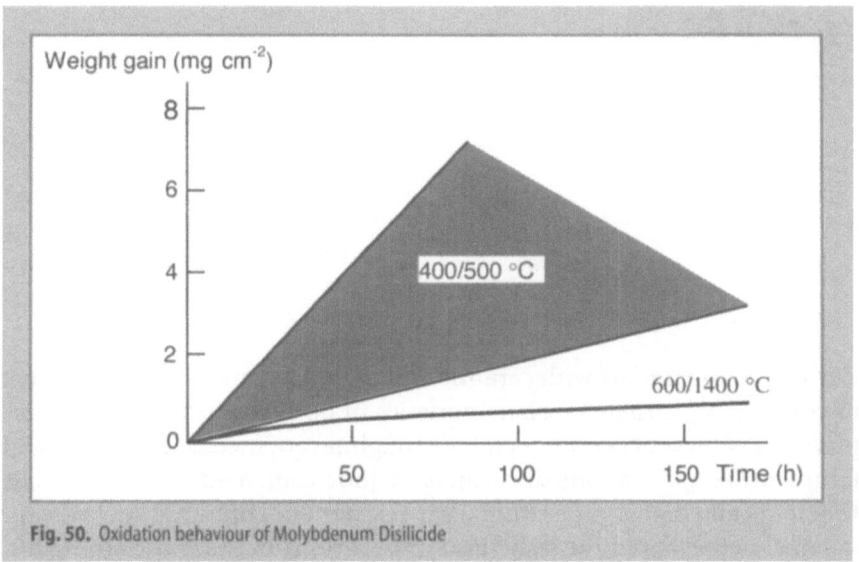

Fig. 50. Oxidation behaviour of Molybdenum Disilicide

been $MoSi_2$, which has demonstrated excellent oxidation resistance up to 1600°C in commercial heating element applications. Protective SiO_2 scales are formed readily at high temperatures. However, as shown in Fig. 50, oxidation rates at 400/500°C are very rapid, and can result in material disintegration – the so called "pest" effect [21]. The effect is not fully understood. It has been studied most extensively for $MoSi_2$ but it occurs in a wide range of intermetallic materials. Other problems with $MoSi_2$ are brittleness at ambient temperatures and poor creep strength above 1200°C. Potential solutions to these problems are being researched and one advantage of $MoSi_2$ is the ability to metallurgically alloy with other silicides to improve properties. Work on the possibility of using $MoSi_2$ as the matrix for high temperature composites has been carried out [22].

Several attempts have been made in recent years to select alternative potentially useful materials on the basis of melting point and simple pragmatic tests for plasticity, oxidation resistance, ductility, etc. Some interesting compounds have been identified but, at this stage, this area of intermetallic materials must be regarded as extremely speculative.

12 Cermets

Cermets are materials with ceramic and metallic constituents, developed to combine the beneficial characteristics of both constituents, for example, the hardness of ceramics and the toughness of metals. The major high temperature applications of cermets include cemented carbides for cutting tools and high temperature wear resistant coatings. Cermets are not normally considered for structural applications because the strength of the metal/ceramic bond is inadequate.

12.1
Cemented Carbide Cutting Tools

The need to increase productivity in manufacturing industry led to a requirement to increase the speed of metal removal in machining with the consequent need for cutting tools better than steel. Severe stresses and temperatures are generated during machining and the ideal tool material has high Young's Modulus to resist deflection, high hardness and strength at elevated temperature to resist deformation, good thermal shock resistance and high chemical stability to minimise reaction with the workpiece.

By 1850 steel with 0.9–1.3% C was the standard cutting tool material. In 1900 an early form of alloy tool steel containing 1.7%C and 9%W was shown to have the capability of running at red heat during machining without losing its edge. This was superseded by the standard 18%W 4%Cr 1%V 0.75%C 18-4-1 high speed steel, which was solution treated at high temperature followed by secondary hardening at 600°C. It was capable of machining at approaching an order of magnitude higher speed than the earlier high carbon tool steels [1].

The first commercial carbide cutting tools were produced in 1927. They consisted of WC particles bonded with small amounts of cobalt, typically around 6%. Crater wear observed on the rake face of the tool was found to be due to solution of WC in the steel workpiece at the high temperatures encountered in machining. Second generation carbide tools contained additions of TiC and TaC which decreased the carbide solubility in steel at 1250°C by an order of magnitude. A typical second generation carbide would be WC-19%TiC-19%TaC-9%Co.

Manufacture is usually by powder metallurgy techniques, with sizing and mixing of powder materials, cold compaction and sintering. Metal coated ceramic particles are sometimes used to assist dispersion of the ceramic in the structure. The early carbide tools were shaped tips, which were brazed to steel shanks. Subsequently higher productivity improvements such as a tool holder with an indexable tip having multiple cutting edges were introduced. Thin coatings (<10μm) of TiC or TiN were subsequently found to further improve tool performance by decreasing the friction coefficient and by significant reduction in chemical reactivity. Such coatings are widely applied to steel and carbide cutting tools. In 1990, coated carbide tools accounted for some 30% of the carbide tool market. The progressive improvement in cutting tool capability is illustrated in Fig. 51 in which cutting speed is related to the machining of medium carbon steel.

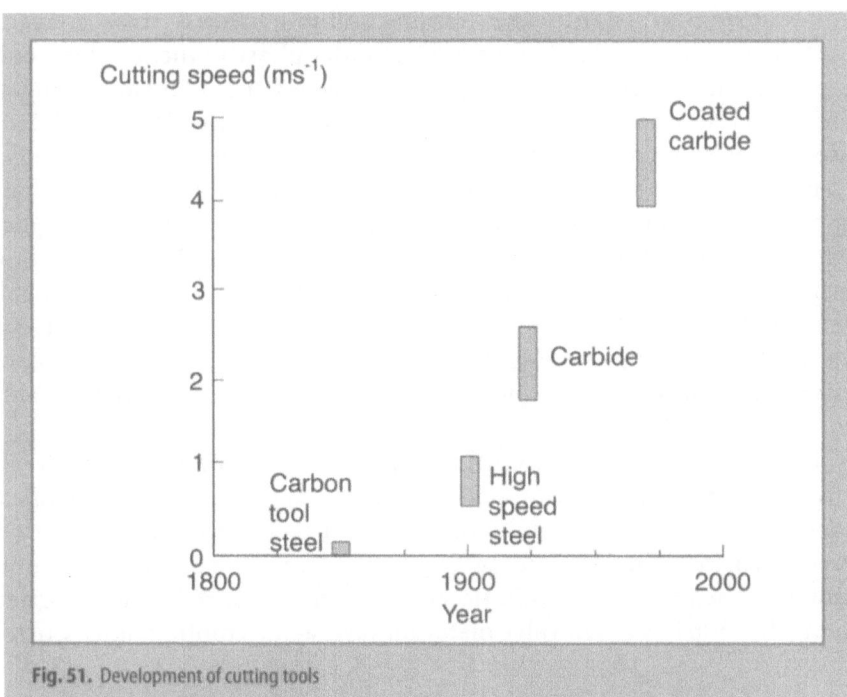

Fig. 51. Development of cutting tools

12.2
Wear Resistant Coatings

While lubricants of various types are commonly used to reduce wear, wear resistant surfaces are often required. Nitriding and carburising have long been used in general engineering. Nitrided surfaces can be used at temperatures approaching 500°C and have increased the life of high-speed steel machine tools by 50% in some applications. Nitriding cannot be classed as a major high temperature wear resistant coating, however. The use of surface treatments and coatings to overcome tribological problems is commonplace. As indicated earlier, wear has been a cause of component damage in many applications, and the type and severity of wear depends on several interacting factors [2]. Selection of wear resistant coatings must take account of operating conditions and the associated need for factors such as hot hardness, toughness, oxidation resistance and thermal fatigue resistance. Other factors important in coating performance are structural stability and lubricants. Coating selection is normally made on the basis of tests developed to simulate the particular environment.

Most coatings contain stable carbide-forming elements (e.g. chromium, tungsten) with hardness increasing and toughness decreasing as the carbide volume fraction increases. The binder phase in the simplest coating is cobalt and it is necessary to select the cobalt content according to the particular wear environment. Thus, in aero gas turbines WC-9Co is used in bearing sealing surface internal diameters while WC-15Co is necessary for fan blade mid-span "clappers" where hammer wear conditions apply. WC-Co can be used at temperatures up to about 500°C. At higher temperatures both WC and the cobalt matrix have inadequate oxidation resistance and improved coatings are required. Many of these are cobalt based to take account of the beneficial wear characteristics of cobalt oxides. All contain high volume fraction of stable carbides and some contain borides in addition. Typical coating compositions are given in Table 20.

The exhaust valve seats in piston engines experience a severe wear environment, being exposed to high temperature, thermal and mechanical shock. Diesel engines running on residual fuels experience the most severe corrosion, with temperatures up to 500°C. Stellite 6 is the most common hardfacing in this application but significant work has been carried out to develop improved valve material/hardfacing combinations. Corro-

Table 20. Wear resistant coating compositions

	C	Cr	W	Ni	Si	V	B	Co
Stellite 6	1	27	5	-	1	-	-	balance
Stellite 12	2	30	10	-	1	-	-	balance
CM 64	0.8	28	19	5	-	1	-	balance
Colmonoy 8	1	26	-	balance	4	-	3.3	-

sion is a less severe problem in petrol engines but temperatures are higher in highly rated engines.

The shrouds of turbine blades, particularly in military jet engines, experience probably the most severe wear environment of all. Temperatures range from 400–1000°C depending on blade stage. Temperature changes introduce a significant thermal fatigue element and the mechanical environment involves some aerofoil twist/untwist under centrifugal load combined with impact loading between adjacent blades. Stellites are used on cooler blade stages while Cr_3C_2-25% NiCr and, more recently, CM64 are necessary on the hotter stages. Hammer wear test data for uncoated superalloys and for various wear resistant coatings is shown in Fig. 52.

It was noted earlier that weld deposition is a standard repair practice in some engineering applications. TIG welding is also used to apply some wear resistant coats such as the Stellite materials. Depending on the coating thickness required, several weld passes might be necessary. Even on relatively thin coats, a two-pass sequence may be necessary because of dilution of the weld metal by the base material on the first pass.

While electroplating [3] has been used to apply wear resistant coats, thermal spraying is the predominant application process. There are several different thermal spray techniques capable of applying good quality coatings from wire, rod or powder sources. Thermal spraying is the generic name for a variety of processes in which material is heated rapidly and projected at high velocity onto a prepared substrate surface [4]. The major processes used in the deposition of wear resistant coatings are flame spray, arc wire spray, detonation gun and plasma spray [5]. In combustion flame spraying an oxygen-acetylene mixture provides the combustion flame and coating powder or wire is fed into the flame, melted

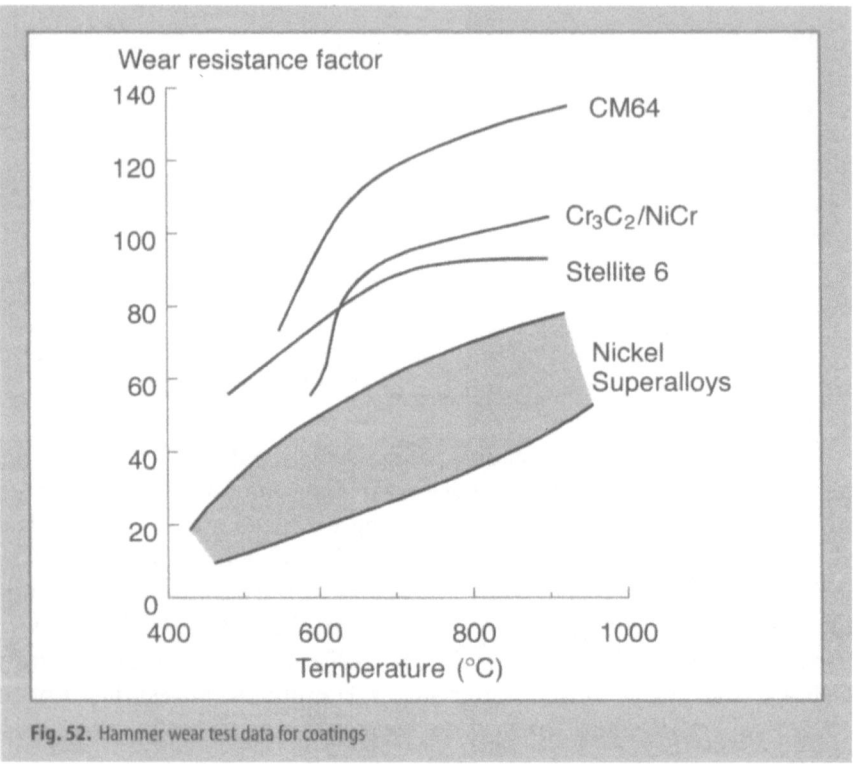

Fig. 52. Hammer wear test data for coatings

and projected onto the substrate. Arc wire spraying produces melting at
an arc between the tips of two consumable wires. A jet of compressed air
atomizes the molten material and projects it onto the substrate surface.
The D-Gun process [6] uses a controlled series of detonations of oxygen-
acetylene mixture to melt the coating powder and propel it through the
gun barrel onto the substrate. In plasma spraying, gases such as argon or
nitrogen are passed through the arc developed by the gun. They are dis-
sociated producing a plasma jet which leaves the gun at high velocity.
Powder material is fed into the plasma jet. The various flame spraying
processes produce different coating characteristics, with flame temperature and particle velocity being important factors. Flame spraying is the
cheapest but the deposit tends to be porous and poorly bonded largely
because of the low particle velocity. Arc sprayed coatings are also relatively cheap and are more dense and better bonded than flame sprayed
coatings. High deposition rates can be achieved. The D-Gun is a higher

Table 21. Characteristics of thermal spray processes

Process	Flame spray	D-Gun	Arc spray	Plasma spray
Flame temp. (°C)	3500	3500	6000	12000
Particle velocity (ms^{-1})	100	800	100	500(air), 800(vacuum)
Relative cost	1	4	2	6 (air)10 (vacuum)

Table 22. Characteristics of WC-Co coatings applied by different thermal spray processes

	Flame spray	D-Gun
Coating-substrate adhesion (MPa)	8	> 70
Porosity (%)	10-15	< 1
Oxide content (%)	10-15	1-5

velocity process and coatings are dense and metallic coatings have a low oxide content. Plasma sprayed metallic coatings tend to have oxide contamination if sprayed in air. Air spraying is widely used for oxide coats, as in thermal barrier coats for example. For metallic coatings, argon shrouds or low pressure spray (LPPS) are required and this increases the cost over other thermal spraying processes. For all processes, the integrity of the deposit is critically dependent on the condition of the substrate surface. Cleanness is of prime importance and all sprayed coatings have significantly higher bond strengths on roughened surfaces [7].

Prime characteristics of the various thermal spray processes are given in Table 21.

While there are some limitations, most of the thermal spray processes can spray most wear resistant coatings. The choice of process usually depends on the duty of the coating. In situations where a measure of impact loading may be involved in service, the highest quality of coatings in terms of density and degree of bonding with substrate together with minimum oxidation is important. This may require D-Gun or plasma spray processes. In applications where this highest quality of coating may not be necessary, flame spraying and arc spraying may be cost-effective possibilities. These process controlled aspects are illustrated in Table 22, which compares the properties of WC-Co coatings applied by flame spraying and D-Gun [5]. Other process selection factors include ability to control heat input into the particular component geometry and ability to automate the process.

With respect to coating material selection, it is possible to spray virtually any material, provided that it melts or becomes substantially molten and suffers no major degradation during the melting/deposition time of the process. Coating material selection thus depends primarily on wear performance in the component environment, a major factor in which is temperature capability.

13 Refractories and Insulating Materials

Refractory materials are used for lining high temperature furnaces and process equipment. Material selection, from the wide range of materials available, depends on factors such as operating temperature, environment, thermal cycle and process economics. The production and properties of refractories are well-documented [1].

Refractories can be categorised into clay based and non-clay-based materials. Refractories manufactured from naturally occurring clays have been used for centuries in applications such as smelting furnace linings and crucibles for glassmaking. These early fireclay materials were based on alumino-silicate clays and contained 25–40% Al_2O_3. The realization that refractoriness increased with Al_2O_3 content led to selection from naturally occurring clays to provide the improved capability required by the increasing demands of the developing iron-making industry in particular. High alumina clays containing 50–80% Al_2O_3 have higher temperature resistance. Fired refractory bricks for lining furnaces and hearths were produced in the early 19th century, initially from fireclays and subsequently, from silica with a lime bond. The development of the basic steelmaking processes later in the 19th century depended on the availability of basic lining refractories such as magnesite and chromite, in which the MgO particles were bonded with pitch or tar.

An alternative to the use of refractory bricks for hot face linings is the use of castable or rammable materials. Use of such materials has increased from the 1920's/30's to the point where currently they are roughly equivalent to bricks in quantity usage. Castable refractories use the same range of mineral types as refractory bricks, but in the form of particulate aggregates bonded together with a cement. The cement varies from a calcium silicate material for low duty applications to a high alumina cement for high temperature uses. The maximum service temperature is dependent in the combined properties of aggregate and cement. A major advantage of castable refractories is the ease with which complex shapes can be produced. Control of water content and the baking process, particularly the drying-out part of the cycle, are essential if excessive cracking is to be avoided. As the firing temperature is increased, sintering develops a ceramic bond in the material. The permanent shrinkage resulting from the initial baking process must be allowed for [2]. It is this shrinkage which leads to the cracking on the hot face of the refractory, and if controlled, this cracking can act like the joints between individual bricks in a brick-built lining and provide an allowance for expansion during subsequent heating.

Refractories, generally, are relatively weak materials and are used in low-loaded applications such as self-supporting linings in kilns and heat treatment furnaces and as supported linings in furnaces and ladles containing molten metal. Typical properties of refractory materials are given in Table 23. Castable materials are weak compared with dense refractory bricks

The thermal protection of a high temperature vessel in a process plant, combustor etc commonly involves the use of several layers of refractory material. The inner 'hot face' layer is exposed to the full process environment, temperature, potentially erosive and corrosive gases/liquids, flowing or static solids, etc. In order to provide the required erosion/corrosion resistance and service performance, this surface layer must be relatively dense. It could be made from bricks or castable/rammable materials. The choice will depend on component shape and de-

Table 23. Properties of typical refractories

Material	Cold crushing strength (MPa)	Thermal conductivity $Wm^{-1}K^{-1}$	Maximum service temperature (°C)
Dense fireclay brick	25 – 60	4	1400 - 1600*
Insulating fireclay brick (60% porosity)	10 – 20	0.5	1400 - 1600
Magnesite brick	35 – 60	6	1600 - 1800**
Clay bonded SiC	80	16	1700

*   Depending on composition
** Depending on binder

Table 24. Maximum service temperatures for insulating castable refractories

Aggregate	Maximum service temperature (°C)
Sillimanite	1600
Perlite	1340
Vermiculite	1100
Pumice	1090

Table 25. Maximum service temperatures for fibrous insulating materials

Fibre material	Maximum service temp (°C)
SiO_2	1250
$Al_2O_3\ SiO_2$	1350
Al_2O_3	1750

sign. However, the dense refractory materials have relatively high thermal conductivity as shown in Table 23 and may not provide adequate thermal insulation. Thus additional backing layers of insulating materials may be used. These layers are not exposed directly to the process environment and therefore can have much lower bulk densities and lower thermal conductivity. Insulating fireclay brick with high porosity is commonly used. Various low-density castable refractories are available – Table 24.

Fibrous insulators are available in the form of blankets and boards, and in a loose form as filler material. They are made from natural or synthetic fibres, with or without chemical bonding materials. Bulk density is a prime factor in determining thermal conductivity, which is minimized at a characteristic packed density [3]. Fibrous insulating materials typically have thermal conductivity in the range 0.2/0.4 $Wm^{-1}K^{-1}$, significantly lower than that of insulating fireclay brick. Temperature capability is determined by the fibre material – Table 25.

Fibrous insulating materials are commonly used in heat treatment furnaces where their lower thermal mass relative to solid insulators allows much shorter heat-up and cool-down times with consequent economic benefits. Lightweight tiles incorporating silica fibres have been used to protect space vehicles during re-entry [4]. These have a surface coating to provide erosion resistance and prevent moisture absorption.

14 Engineering Ceramics

Ceramics are inorganic non-metallic materials including oxides, borides, carbides, nitrides and silicides. Major application areas are electronics and engineering. It is engineering ceramics, used for mechanical applications, often involving high temperature, which are discussed here. The major engineering ceramics are Al_2O_3, ZrO_2, SiC and Si_3N_4.

In addition to their use in structural components, ceramics are also used as insulating materials and in 'functional' applications such as filters. High porosity ceramic foams have been developed for liquid metal filtration applications and for the filtration of particle-containing gases in combustion and gasification power generation.

Early engineering ceramic applications, such as Al_2O_3 for spark plug insulators and SiC for electric resistance heating elements, involved relatively low stress and moderate to high temperature. Somewhat elementary material behavioural understanding allowed improvement in strength and temperature capability. During the last 25 years, the understanding of the relationship between processing, microstructure and properties has improved significantly. Much work has been carried out on high temperature applications involving higher stress levels and progress has been made in improving defect tolerance.

14.1
Manufacture

Unlike the clay based refractories, engineering ceramics are made from refined natural or synthetic minerals. Material in powder form is the normal starting point in manufacture. Powder purity, size distribution and mixing are of major importance in the early process stages. Additives include binders to provide strength during green body handling and machining, plasticisers and wetting agents as aids in shaping operations and sintering aids. The fabrication process normally consists of three main stages: production of a green "shaped" component from the powder; densification by high temperature firing; finishing to the required tolerances.

The major shaping processes are die pressing and isostatic pressing for relatively simple shapes, machining green bodies, injection moulding and slip casting for more complex shapes. In die pressing, the powders are compacted in a die under uniaxial pressure or under isostatic pressure in a compliant mould. Ceramic/plastic fabrication techniques involve mixing the ceramic powder with binders and plasticisers to aid formability. Injection moulding is probably the best technique for complex shapes, with extrusion and rolling processes being used for simpler forms. Slip casting is a well established process in which the shaped component is produced from a powder suspension in a porous mould.

Densification processes involve some form of sintering. Sintering involving chemical reaction is used for reaction bonded silicon nitride and silicon carbide. Solid state sintering, usually with added impurities to modify surface energy and reactivity is possible – for example SiC may be sintered with boron and carbon sintering catalysts. Totally solid phase sintering is rare however and for most systems diffusion kinetics require the presence of a liquid phase during sintering. This necessitates the addition of appropriate sintering aid powders to the ceramic powder prior to the shaping process. Significant shrinkage occurs during sintering and residual porosity is minimized in hot pressing and hot isostatic pressing which apply temperature and pressure simultaneously.

The need for final finishing to required tolerances must be minimised as far as possible because ceramic materials are difficult to machine due to their high hardness. Indeed, some ceramics are used as tools to machine metals. The emphasis in shaping/densification is thus on near-net shape fabrication. Because of the possibility of introducing cracks during machining due to the brittleness of ceramics, carefully controlled machining schedules are necessary. Mechanical, thermal and chemical techniques are used in the removal of ceramic material.

All aspects of ceramic manufacture are well-documented [1].

14.2
Properties

The strength of various ceramic materials is shown in Fig. 53, with a typical precipitation strengthened nickel superalloy included for comparison.

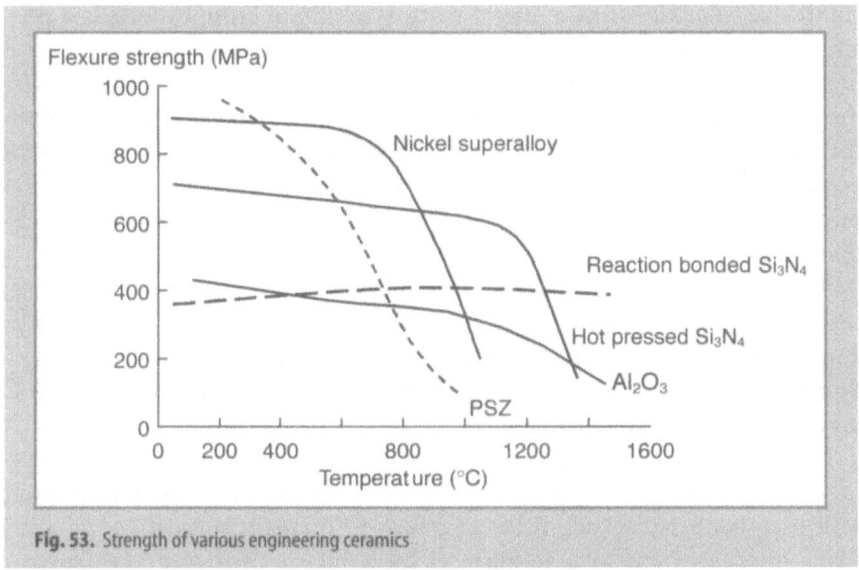

Fig. 53. Strength of various engineering ceramics

The higher temperature capability of silicon nitride and silicon carbide over nickel superalloys, together with much lower density, has provided the motivation for the interest in ceramics in several major gas turbine programmes.

However, as was noted earlier, the major problem with monolithic ceramic materials is poor defect tolerance arising from fundamental lack of dislocation movement. The tendency of ceramics to fail catastrophically by growth of a single crack originating from a very small defect, because of low fracture toughness, limits their reliability. The critical defect size may be in the range 1μm–100μm in diameter, which is much smaller than that for metals under similar stressing. Because of their small size, it is difficult to avoid these critical defects in manufacture and to detect them by inspection.

Three approaches have been adopted in attempts to improve the reliability of ceramics [2]. The first is to develop a fracture mechanics based probabilistic approach using failure micromechanics (the relationship between defects and strength [3]) in order to specify allowable design stresses and life. In practice however, it is difficult to specify the flaw characteristics in the actual components sufficiently accurately by statistical strength characterisation. It is also difficult to specify the flaw characteristics by non-destructive evaluation (NDE) because of the small size of critical flaws. Proof testing of actual components is a possible alternative approach for ensuring a specified strength. However it is not completely satisfactory because the proof test itself can modify the flaw population [4]. The major problem with the probabilistic fracture mechanics – based approach however is that, even if the initial flaw population could be specified sufficiently accurately, new flaw populations could be created during service which would invalidate the predictions.

The second approach is to develop new manufacturing processes which will result in reduced flaw size thus allowing design stresses to be kept below the failure stress. Advances have been made in process development but the possibility of new flaws being created in service tends to invalidate this approach also.

Increasing inherent defect tolerance by designing ceramic microstructures with improved fracture resistance is the most realistic approach. This has been achieved in several ways. Transformation toughening Al_2O_3 increases fracture toughness by a factor of 4 and the fracture toughness of Si_3N_4 can be doubled by control of microstructure allied to sintering. Increasing interest has been shown during the last decade in ceramic composites and ceramics reinforced by continuous fibres are showing good toughness [5]. Increasing the inherent defect tolerance is especially important where a hostile operating environment can introduce additional strength – degrading defects.

The ability to withstand rapid changes in temperature is a requirement in some ceramic applications, for example the potential replacement of nickel superalloys in gas turbines. As was noted earlier the thermal shock resistance of ceramics is strongly dependent on physical properties, with Si_3N_4 and SiC having significantly better thermal shock resistance than Al_2O_3.

14.3
Alumina

Alumina is probably the most widely used engineering ceramic. A wide range of Al_2O_3 ceramics is commercially available with strength and temperature capability depending on Al_2O_3 content, which is normally in the range of 85 to 99%. The high hardness of Al_2O_3 makes it difficult to machine sintered material and presinter shaping capability decreases with increasing Al_2O_3 content.

Table 26. Performance of Alumina-based cutting tools

Material	Maximum cutting speed (surface ms^{-1})
Tool steel/high speed steel	1.2
Carbides (WC-Co)	2.5
Al_2O_3/Al_2O_3 - TiC	5
Transformation toughened Al_2O_3	12.5
Al_2O_3/SiC_w	

Alumina is widely used for crucibles, thermocouple sheaths and rods for general high temperature applications. Perhaps the most important single product is spark plug insulation but use of Al_2O_3 in cutting tools is increasing. Compared with tool steels and sintered carbide tips, ceramic cutting tools do not deform at high temperature and have superior chemical resistance at workpiece/tool interface conditions. They are thus capable of faster cutting speeds. As noted earlier the performance of Al_2O_3 tools has been improved in recent years by the addition first, of TiC particles, and subsequently SiC whiskers. The fracture toughness of Al_2O_3 is increased by a factor of 4 by the incorporation of SiC whiskers. Unlike transformation toughened Al_2O_3, the toughening mechanism in Al_2O_3/SiC material is not temperature limited and it is used for rough machining at higher cutting speeds than is possible with carbide tips –Table 26. This is partly due to the increased defect tolerance and partly due to the fact that the silicon carbide whiskers increase the thermal conductivity and consequently the thermal shock resistance. The Al_2O_3/SiC material has a fourfold increase in cost over carbide tips but a tenfold increase in life. It cannot be used to machine cast iron at high speed however, because of the reaction of SiC with iron.

14.4
Zirconia

Apart from its high melting point, zirconia has two important characteristics. It has low thermal conductivity and it is polymorphic. The monoclinic phase is stable to 1170°C, the tetragonal phase from 1170 to 2370°C and the cubic phase is stable from 2370°C to the melting point.

The tetragonal-monoclinic phase transformation involves a volumetric change of around 4%, which can cause failure on cooling large pieces of pure ZrO_2. Phase transformation must therefore be prevented in some situations and when ZrO_2 is used in applications such as extrusion dies, atomising nozzles and melting crucibles, the cubic phase is normally stabilized by the addition of CaO. The phase transformation can however be used to advantage to improve the toughness of ceramics. It was first applied to ZrO_2 in 1975 and has subsequently been applied to other ceramics. The principle, known as transformation toughening [6, 7], makes use of the volumetric expansion which occurs when the tetragonal phase transforms to the monoclinic phase. In the toughening of ZrO_2 itself, small particles of oxides such as CaO, MgO and Y_2O_3 are added at the powder mixing stage. The quantities used are insufficient to completely stabilise the cubic phase and heat treatment at 1400°C then produces small spheroidal precipitates of tetragonal ZrO_2 in a cubic matrix. For other ceramic materials, ZrO_2 particles of a critical size are introduced into the ceramic at the powder stage. It is generally accepted that there may be two different mechanisms involved in the toughening – micro-crack toughening and stress-induced transformation toughening. In micro-crack toughening, illustrated in Fig. 54, the expansion associated with the transformation of the tetragonal particles to the monoclinic

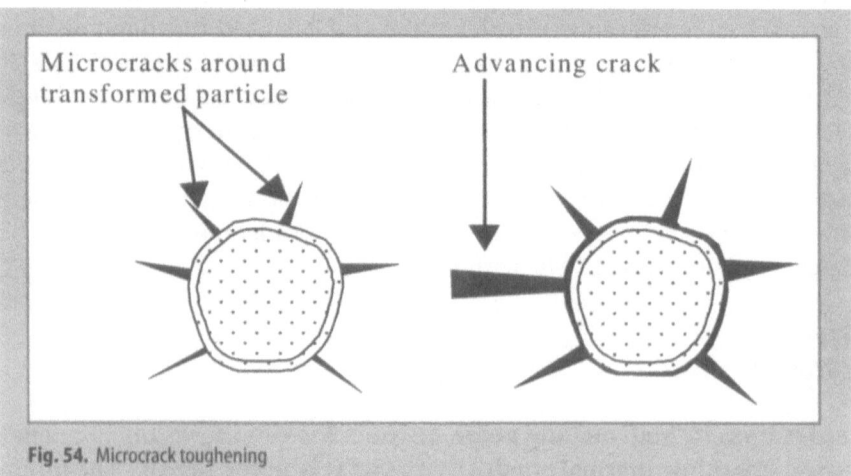

Fig. 54. Microcrack toughening

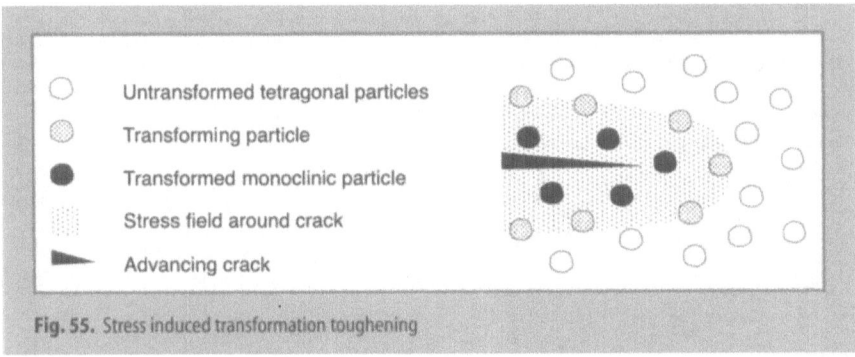

Fig. 55. Stress induced transformation toughening

phase causes sub-critical micro cracks to form around the particle on cooling from firing temperature. These then deflect and branch a crack propagating into the particle, thus lowering the energy of the crack. In stress induced transformation toughening, an advancing crack releases the matrix constraint which had prevented the tetragonal/monoclinic phase transformation and allows the monoclinic phase to form in the material around the crack tip. This introduces a local compressive stress at the crack tip thus retarding crack propagation – Fig. 55. Transformation toughened zirconia, sometimes referred to as partially stabilized zirconia (PSZ) is used for cutting and slitting industrial materials.

Zirconia has one of the lowest thermal conductivities of ceramics, in the dense form – Fig.56. The thermal insulating capability of ZrO_2 has been utilized in diesel engines and gas turbine engines mainly in the form of thermal barrier coatings (TBC) to date.

14.5
Silicon Carbide

Silicon carbide is widely used as rods or tubes for electric resistance heating elements, in high temperature kiln furnaces and heat recovery equipment. Depending on application, the material may be clay bonded or self-bonded. Clay bonded material is relatively low duty, low cost. Reaction bonded material is produced by infiltrating molten silicon into a powder compact of SiC and carbon. The silicon reacts with the carbon to bond the structure together but the final material contains 10–20% residual silicon, which causes a sharp fall in strength at temperatures around 1200°C (Fig. 57).

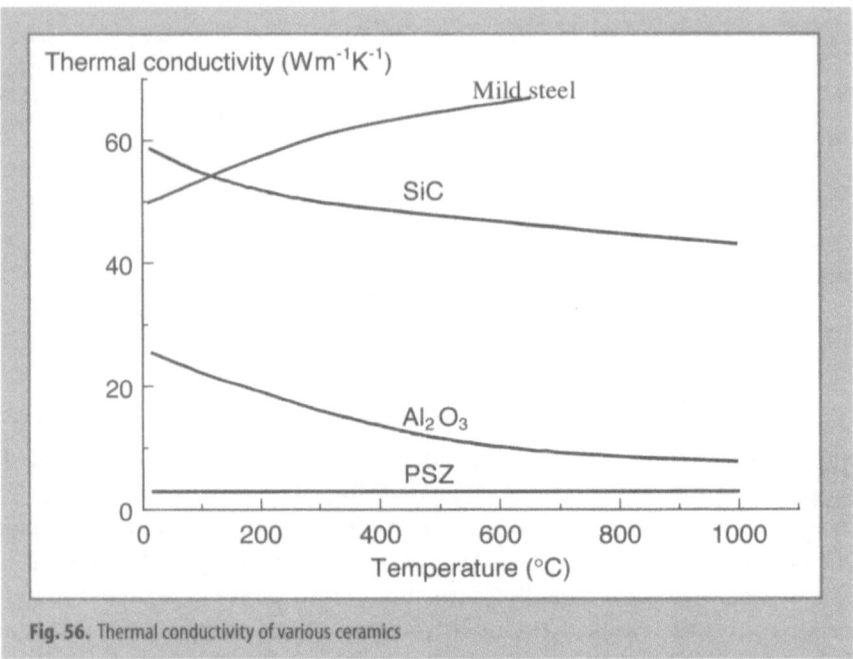

Fig. 56. Thermal conductivity of various ceramics

The maximum service temperature of heating element material which is produced by pressure sintering bonded powder and firing at high temperature depends on life requirement. For long life the maximum temperature is around 1450°C with a protective film of SiO_2 forming in oxygen at temperatures above about 1100°C.

The major automotive ceramic gas turbine programmes in the 1980's used components manufactured by slip casting, injection moulding and hot pressing. SiC can be solid state sintered at temperatures of 1900–2100°C with small additions of carbon, probably acting as a surface deoxidizing agent, and boron to modify the surface energy and diffusivity. Because of the absence of grain boundary phases, strength is maintained to high temperature -Fig. 57. Sintering can be carried out at lower temperatures under pressure with boron and Al_2O_3 sintering aids and hot pressing uses some 2% Al_2O_3 to aid sintering.

Ceramic heat exchangers have major potential in fuel conservation by utilising waste heat recovery in a variety of situations including industrial furnaces, industrial cogeneration and fluidised bed combustion. It has been estimated that fuel savings in excess of 60% are possible in certain

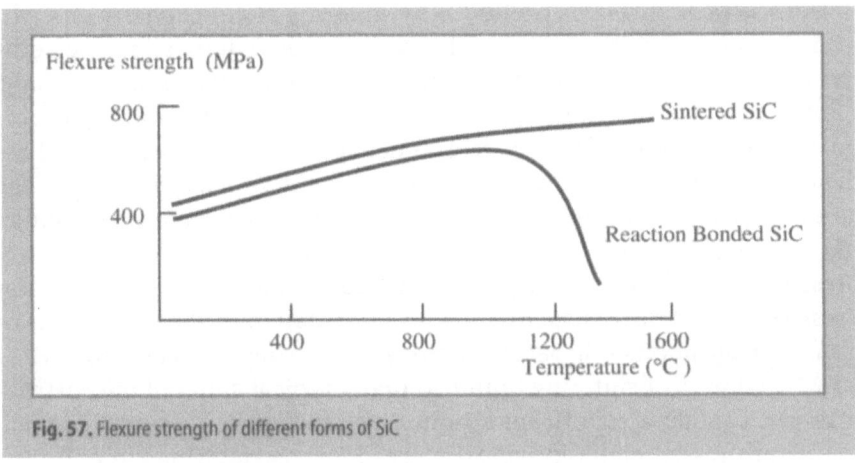

Fig. 57. Flexure strength of different forms of SiC

applications [8]. Sintered SiC and various aluminosilicates have been used in low-pressure heat exchangers with component feasibility demonstrations at higher temperatures [9].

In electric power generation, the objectives of future fossil-fuel power plants are to minimise the generating cost and meet emissions legislation. Fluidised bed and coal gasification hybrid concepts involving gas turbines are potential commercial plant of the future. In both cases it is necessary to remove particulate material from the combustion gas to avoid erosion problems in the gas turbine [10]. Filtration using porous ceramics appears to be the preferred technology [11] and clay bonded SiC porous "candles" have been used in some programmes.

14.6
Silicon Nitride

Silicon nitride is produced by reaction bonding (RBSN), by sintering and by hot pressing. RBSN is produced by manufacturing a silicon powder compact, shaping this in the pressed "green" state and nitriding in a nitrogen atmosphere at a temperature of 1400°C. There is very little shrinkage during nitridation so that the machined green shape can be close to final dimensions. However the material has a significant porosity level. Thus the strength is relatively low but because the material is essentially additive free, is maintained to high temperature. (Fig. 53)

Dense silicon nitride is produced by sintering at atmospheric or slight over pressure or by hot pressing. Both processes require sintering additives to achieve liquid-assisted sintering at temperatures in the range 1700–1800°C because solid state diffusivity is too low below the decomposition temperature [12]. Inter-granular phases which are formed reduce high temperature properties to an extent which depends on the amount and composition of inter-granular phase. The amount of sintering aid required generally decreases with applied pressure, so that hot pressing may only involve transient-liquid sintering. A typical hot pressed (HPSN) microstructure may thus contain only small amounts of glass at triple points -Fig 58a. Pressureless sintering normally requires 5-10% by volume of sintering liquid so that a typical sintered microstructure will contain a significant amount of liquid phase residue -Fig 58b [13]. Hot pressed material typically uses MgO as sintering aid and thus contains a glass of low viscosity because of reaction between the MgO sintering aid and SiO_2 from surface contamination of the Si_3N_4 starting powder. Commercial brands of sintered material (SSN) use various additives – MgO, $MgO+Al_2O_3$, ZrO_2, $Y_2O_3+Al_2O_3$ etc. There is often a compromise between sinterability and the softening temperature of the glass residues and consequently high temperature strength. It is possible to select additives such that the liquid residue can be crystallized by a post-sintering heat treatment to give further improvement in high temperature creep strength – Fig 23.

The morphology of the β Si_3N_4 crystal growth can be controlled to some extent and can lead to a toughening increment. Fracture toughness values for the essentially mono-phase HPSN material are around 4 $MPam^{1/2}$ and for the bi-phase SSN material are around 7 $MPam^{1/2}$. This is because toughening mechanisms such as grain pull out and crack deflection operate [13].

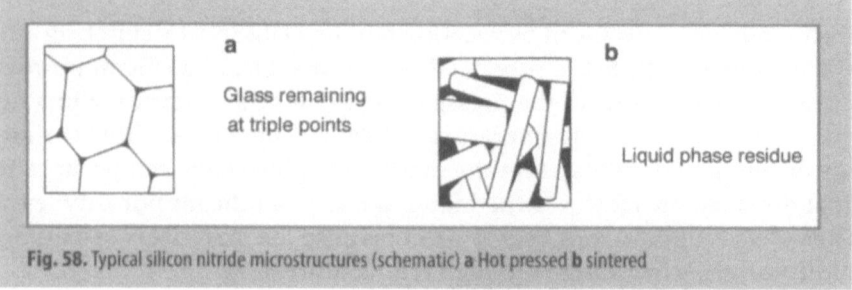

Fig. 58. Typical silicon nitride microstructures (schematic) **a** Hot pressed **b** sintered

Pure silicon nitride has good oxidation resistance due to the parabolic oxidation kinetics of the protective SiO_2 surface layer. The presence of a second phase can severely degrade the oxidation resistance by three possible mechanisms:
- stresses due to volume changes because of second phase oxidation
- reaction of the second phase with SiO_2 to produce a low temperature glass
- diffusion of reacting species through the glassy phase

A post-fabrication oxidation treatment can significantly improve oxidation resistance [12].

Si_3N_4 has been used in ceramic automotive turbine programmes for turbine rotors and vanes, scrolls, combustor baffles etc. [14, 15].

The incentives for the use of ceramics are the potential for:
- Increase of 200 °C in turbine inlet temperature for uncooled designs, compared with superalloys
- Consequent increase in thermal efficiency from 40 to 50%
- Reduced engine size and weight
- Reduced exhaust emissions

Power increases of 40% and fuel savings of around 10% have been demonstrated in research engines containing ceramic components [16]. While it has been claimed that proof of concept has been established, lack of reliability and defect tolerance has nevertheless proved to be a major ongoing problem with monolithic ceramic materials.

Silicon nitride has also been used in demonstrator adiabatic diesel engines for pistons, piston caps, cylinder head shields and for turbocharger rotors in production automotive engines. Application to other automotive engine components could be close. The driving forces include emphasis on fuel economy and environmental concerns. The lower weight and higher temperature capability, compared to metals, potentially could result in reduced emissions and improved fuel economy with no detriment to power [17]. Reliability and cost have slowed market acceptance. A study of the economics of replacing certain metallic automobile valvetrain components with ceramic, in new designs, concluded that the horizon for commercial use of ceramics could be some years away, but is worth pursuing [18].

Because of its hardness retention at temperature, Si_3N_4 has recently been introduced as a cutting tool for the rapid machining of cast iron.

14.7
Glass Ceramics

The thermal expansion behaviour of glasses is controlled by composition and by the thermal history of the total manufacturing process. Fused SiO_2 has a very low coefficient of thermal expansion and consequently has excellent thermal shock resistance, reflected in its industrial and domestic uses.

The concept of producing polycrystalline materials from glass is not new. Developments have led to combining the benefits of glass shaping processing (ability to produce relatively complex shapes with good dimensional accuracy) with the property improvement resulting from devitrification –the conversion of glass to a fine grained polycrystalline solid. The manufacturing process involves smelting and shaping a glass of selected composition containing nucleating agents for crystallization. A duplex devitrification heat treatment is applied [19]. Tiny crystalline nuclei are formed in a lower temperature heat treatment and then grown in a higher temperature treatment. Devitrification may occur in glass during use, as in the formation of cristobalite in fused SiO_2 in long term use at temperatures above 1000°C.

The glass ceramic lithium-aluminum silicate (LAS) has a low coefficient of thermal expansion and excellent thermal shock resistance – (Table 4). As noted, the use of heat exchangers to recover heat in industrial processes is increasing in importance. Thermal shock resistance is a key requirement because of the temperature gradients existing during operation. LAS regenerator cores have been reported to have run for over 10.000 hours in truck gas turbine engines [20]. LAS is used for domestic cooker hobs and magnesium aluminum silicate (MAS) has been used for missile radomes.

15 High Temperature Composite Materials

The emphasis to date in the development of high performance composite materials has been to improve specific properties within the temperature capability of the existing systems rather than to increase the temperature capability. Historically high performance composite materials have been developed primarily to meet the aerospace industry need for strong, lightweight structures of high stiffness. The glass fibres, which were available as reinforcements up to 1960, were strong but not particularly stiff. The production of carbon fibres in the early 1960's led to high-strength/high-stiffness polymer composite structures in which the properties were controlled by varying the amount and stacking of the fibres in various directions. Such materials are widely used in the latest designs of airframe (some 10% and 30% of the structural weight of commercial and military aircraft respectively).

With the manufacture of stiff boron fibres in the early 1960's it was possible to consider the reinforcement of metal matrices, and jet engine fan blades in aluminium reinforced with boron fibres were engine tested in the late 1960's. As far as most aerospace applications are concerned, however, aluminium metal matrix composites (Al MMC's) with continuous fibre reinforcement are not cost effective relative to competing materials. This is partly due to the fibre cost and partly due to cost of component manufacture. Diffusion bonding continuous boron fibres in aluminium structures is some two orders of magnitude more costly than squeeze casting with discontinuous fibres. Aluminium MMC's with discontinuous reinforcement such as chopped Saffil fibre [1] are being used in automotive applications such as the local selective reinforcement of hot-running components, essentially to improve thermal fatigue resistance [2].

A number of fibres were evaluated as possible reinforcements for titanium in the early 1960's, including alumina and boron. Excessive fibre/matrix reaction and fibre degradation resulted in poor composite properties, indicating the need for a more stable fibre. Boron fibre was coated with SiC in the first attempt to achieve this, leading eventually to the manufacture of SiC fibres in the late 1970's. This fibre, with subsequent surface modifications, is used in TiMMC materials initially aimed at the US NASP aircraft project and currently being used in the latest jet engine demonstrator programmes.

The development of carbon-carbon composite materials began in the late 1950's. The prime requirement was improved toughness compared with the monolithic graphite material previously used in rocket applications. A wide range of properties is available, depending on fibre type and architecture, composite manufacture and heat treatment, with properties being maintained to high temperature.

In a similar way, the development of ceramic matrix composites (CMC's) had its origins in the need to retain the best properties of ceramic materials together with the additional quality of defect tolerance. Work in the 1960's indicated that carbon fibres could be successfully incorporated into glass matrices to produce high strength, tough composites – Fig. 25. Lack of oxidation resistance limits such composites to proof-of-concept model materials, but the availability in the 1970's of the organo-metallic derived silicon carbide fibres such as Nicalon permitted the creation of composites with improved oxidation resistance. Such composites were first introduced into nozzle flaps in a jet engine in a fighter

Table 27. Properties of some reinforcement fibre materials

Fibre	Strength (GPa)	E (GPa)	Density (gm/cm^3)	Diameter (μm)
E glass	3.0	70	2.5	10
W-1% ThO$_2$	1.0	-	19	200
Boron	2.8	420	2.7	140
Alumina (FP)	1.4	380	3.9	20
Carbon (high E)	2.5	490	1.9	8
Carbon (high strength)	5.0	290	1.9	8
SiC (SCS 6 monofilament)	3.4	430	3.0	140
SiC (Nicalon)	2.8	190	2.6	15
SiTiCO (Tyranno)	2.8	190	2.5	10

aircraft in 1989. Much work is ongoing to improve all aspects of the technology of CMC materials to widen application potential.

The major properties of the fibres that have made possible the development of the high-temperature composite material systems to their present status are given in Table 27.

15.1
Metal Matrix Composites

Work on metal matrix composites (MMC) dates from the 1960's, with early emphasis on aluminium, copper and nickel superalloy matrices. Aluminium MMC's, like aluminium alloys, do not have a sufficiently high temperature capability to be included in this review. Nevertheless, aluminium MMC's can be used to illustrate the relationship between technology benefits and material cost. Where maximum materials capability is essential, as in space technology for weight saving, the high cost of continuously reinforced MMC may be justified. In the space shuttle much of the cargo bay structure is supported on B-Al tubes giving a 44% weight saving relative to aluminium alloys. In contrast, the local discontinuous reinforcement in aluminium pistons in vehicular diesel engines was justified because the short fibre reinforced preform replaced the more expensive Ni-Resist cast iron ring insert [3].

In the early days, refractory metal wires received a great deal of attention as possible reinforcement materials for high temperature composites. Tungsten and molybdenum wires were available as lamp filament and thermocouple wire respectively. Copper/tungsten wire composite was studied, initially as a model system, but ultimately as a possible heat exchanger material for hypersonic aircraft engines. Nickel superalloy/ tungsten wire composites were evaluated for possible rocket combustion chamber and nozzle applications with more recent work being carried out on FeCrAlY/tungsten wire materials for possible turbine blade application. In the end, no real use has been made of these types of material, because of several problems, notably the poor oxidation resistance and high density of the reinforcement.

Various potential applications of MMC in aerospace are discussed [4]. Titanium MMC materials appear to have most promise.

15.2
Titanium Matrix Composites

As was noted earlier, significant progress with TiMMC was not achieved until silicon carbide monofilament fibres became available in the late 1970's. These fibres were produced by depositing SiC by chemical vapour deposition onto tungsten or carbon cores. This produces relatively large 140 μm diameter fibres [5, 6]. Excessive fibre/matrix reaction occurred during composite fabrication by vacuum hot pressing of early fibres. The manufacturing route involved the stacking of alternate layers of aligned, spaced monofilaments and titanium alloy foil, sealing in a can and hot pressing at temperatures around 900/950°C [7]. Ti6-4 is the most commonly used foil material to date. Alternative methods of applying the matrix material have included plasma spray and vapour phase coating processes [8].

A major task in the development of titanium MMC technology is avoiding the formation of the harmful fibre/matrix interaction zone during composite fabrication and service. A brittle reaction product forms at the interface because of the high reactivity between stoichiometric fibres and the titanium matrix. The high mismatch in coefficient of thermal expansion (CTE) between fibre and matrix results in thermal stresses at the interface and, when subject to additional mechanical loading, the brittle interface layer cracks. Modified fibre surface chemistry or fibre coatings have been adopted to act as a diffusion barrier to control the interaction. The first approach involved the creation of a silicon rich outer layer on the fibre to form a reaction barrier together with a graphite carbon coating to act as a compliant layer [9, 10]. Duplex carbon and TiB_2 coatings have also been investigated.

Significant increases in strength and stiffness over unreinforced material are achieved in the direction of reinforcement as illustrated in Fig. 59. In the transverse direction E is increased to some extent over conventional material while the strength is lower than conventional Ti 6-4.

Studies have been carried out to establish the potential of Ti MMC in reducing the weight of jet engine fan blades and increasing the stiffness and reducing the weight of shafts. However the maximum potential appears to be in compressor components in future jet engine designs. For example, compressor discs can be replaced with "blings" – rings onto which the blades are mounted – Fig. 60, giving claimed weight savings of up to 75% [11]. The validity of the concept is dependent on the ring material having the strength and stiffness of Ti MMC.

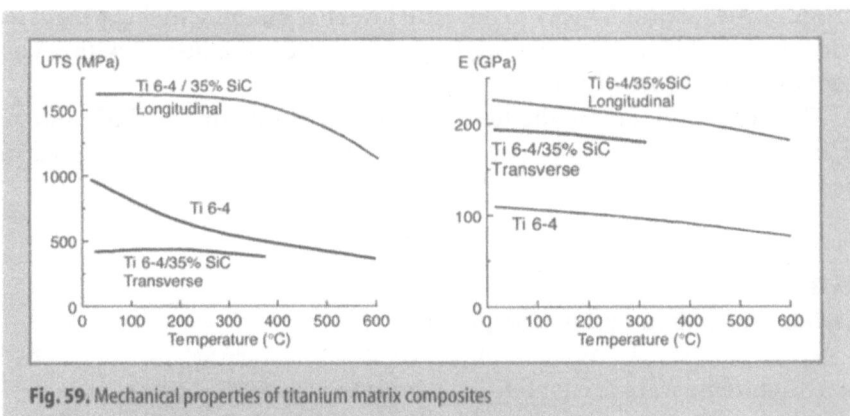

Fig. 59. Mechanical properties of titanium matrix composites

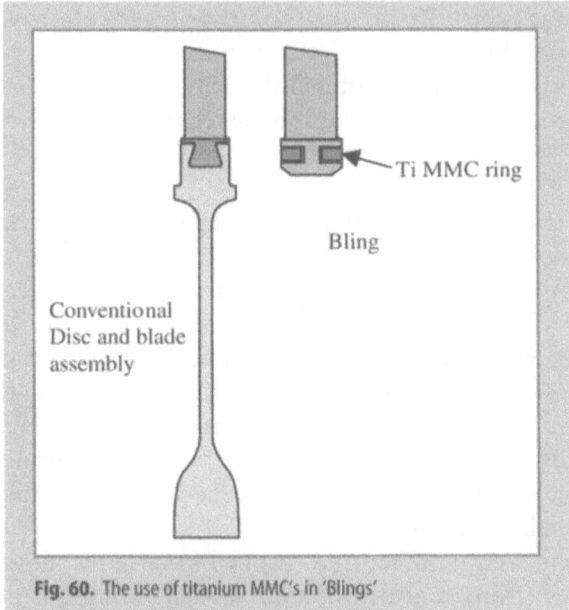

Fig. 60. The use of titanium MMC's in 'Blings'

Engine demonstrator programmes have been run with Ti MMC "blings" and, while the advantages of Ti MMC materials are closer to exploitation, further work is required in several areas. Knowledge of material behaviour must be extended, with predictive failure models being developed and validated. Manufacturing cost reduction is the key issue, however, if these materials are to be established in production quantities.

Programmes are under way to develop low cost manufacturing processes with the capability of producing large quantities of material with a cost goal of around 20% of current cost.

The desire to increase the temperature capability of Ti MMC systems is motivating work on the use of titanium aluminide intermetallics as possible matrix materials.

15.3
Carbon and Carbon-Carbon Composites

The high temperature capabilities of various forms of carbon have been long known, being first exploited in the use of graphite crucibles for melting alloys in the fifteenth century. Subsequently carbon has been widely used in furnaces of various types. Pyrolytic graphite is a very effective thermal insulation material, but inserts in rocket nozzles have experienced brittle failures. This brittleness together with poor oxidation resistance is being addressed in carbon-carbon composite technology.

Carbon
Carbon is available in a variety of forms ranging from amorphous carbon to crystalline pyrolitic graphite.

Carbon has been widely used as blocks for lining furnaces and in electrodes for electric arc furnaces. The higher electrical conductivity of graphite allows electrodes with much smaller cross sectional area than carbon electrodes, for the same current carrying capability [12]. Graphite is also widely used in resistance heating elements where its low vapour pressure allows it to be used up to 3000°C in inert or reducing atmospheres and 2500°C in vacuum. The susceptors in induction heating furnaces are commonly made in graphite.

Shaped carbon products are manufactured from carbonaceous material such as coke particles (from coal or oil, sources depending on the chemical purity required in the final product) combined with a binder phase, usually a coal tar pitch. Green products are formed by pressing in moulds or by extrusion. They are then baked to drive off the volatiles and carbonised at higher temperatures, around 1200°C to give the properties which are required for the particular application. These may emphasise strength, electrical conductivity, porosity, etc. Graphite is made by a similar process, using petroleum coke, with a final graphitisation treatment at temperatures around 2700°C, to give a structure of micro-crystalline graphite particles in a matrix of amorphous carbon.

The low elastic modulus and thermal expansion coefficient and high thermal conductivity of carbon lead to good resistance to thermal shock. Pyrolytic graphite is highly anisotropic, with thermal conductivity in the direction perpendicular to the basal plane of the crystallite being some two orders of magnitude lower than that in the direction parallel to it. Thus it is a very effective thermal insulation material which has led to its use as inserts in rocket nozzles. Brittle failures have been experienced in these components however, indicating the need for tougher material.

Carbon-Carbon Composites

The development of carbon-carbon materials began in the late 1950's, motivated largely by the space shuttle programme in the U.S. The objectives included sufficient strength and stiffness to resist flight loads and thermal gradients, low coefficient of thermal expansion to minimize thermal stresses, oxidation resistance to limit strength reduction and tolerance to impact damage [13].

Carbon-carbon composites consist of carbon fibres arranged in uni- or multi-directional architectures in a carbonaceous matrix. A typical manufacturing sequence is illustrated in Fig. 61.

The carbon fibres are usually made from polymers such as rayon or polyacrylonitrile (PAN). The polymer fibre is heated to around 250°C followed by carbonisation at temperatures above 1000°C and graphitization above 2000°C. High modulus fibres can be produced by carrying out the final heat treatment under tension. Fibre strength or modulus may be optimized (Table 27) by selection of particular processing conditions [13].

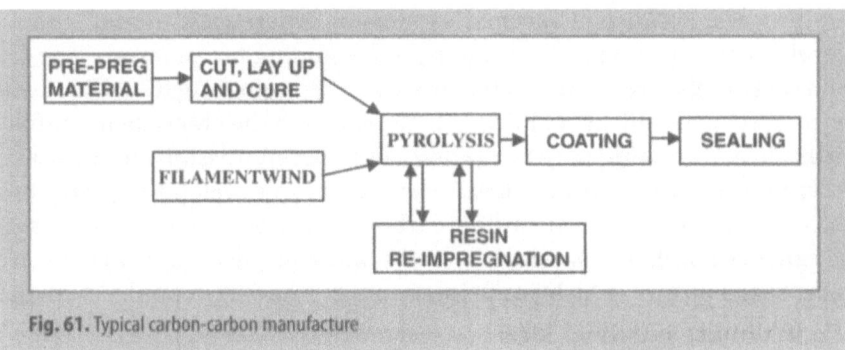

Fig. 61. Typical carbon-carbon manufacture

The fibre architecture can be created by filament winding, weaving, knitting or braiding [14].Carbon is deposited as the matrix within the structure by pyrolysing an organic precursor or by chemical vapor infiltration (CVI) from a hydrocarbon gas at temperatures around 1050/1100°C [15]. Multiple re-impregnation and pyrolyzations will be required in order to achieve the required degree of densification. Further heat treatment at graphitisation temperatures may be necessary to give the required thermal stability and thermal conductivity.

Properties can be varied depending on carbon fibre selection, fibre architecture and process conditions. The strength range of 3D composites is compared with un-reinforced graphite and a typical high strength cast nickel superalloy in Fig. 62. The strength of carbon-carbon composites is reduced partly by their high inherent porosity and partly by the low failure strain of the carbon matrix. Nevertheless, the strength is maintained to temperatures above 2000°C. Toughness varies according to process conditions but is good, with work of fracture in the range 60-125 KJ/m^2.

Carbon-carbon composite materials can withstand temperatures around 3000°C in an inert atmosphere, but the oxidation resistance is similar to that of unreinforced graphite. For single mission applications, such as rocket nozzle or cruise missile engine components this may not be a significant problem. A similar situation applies to multi- mission applications in which the material experiences only brief exposure to high temperature and in which design can restrict air access, such as aircraft brakes. However, for long term use at temperatures above 500°C, oxidation degradation would eventually lead to failure and protective coatings are required.

The carbon-carbon leading edges and nose cap of the space shuttle have a two stage protective system. The first stage involves converting the surface material to silicon carbide by a high temperature diffusion coating process. Because of thermal expansion differences, the SiC coating develops microcracks on cooling from the coating temperature. The second stage in the protective system involves impregnating the surface with tetra-ethylortho silicate (TEOS), which results in the cracks being infilled with silica. Recoating may be carried out between missions if necessary.

From the early 1980's work was carried out to develop long-term protection to allow the use of carbon-carbon in jet engine components such as exhaust nozzle flaps, reheat systems and combustors. Protective systems based on SiC or Si$_3$N$_4$ as primary oxygen barriers coupled with matrix inhibitors and outer layers to seal cracks were evaluated [16]. How-

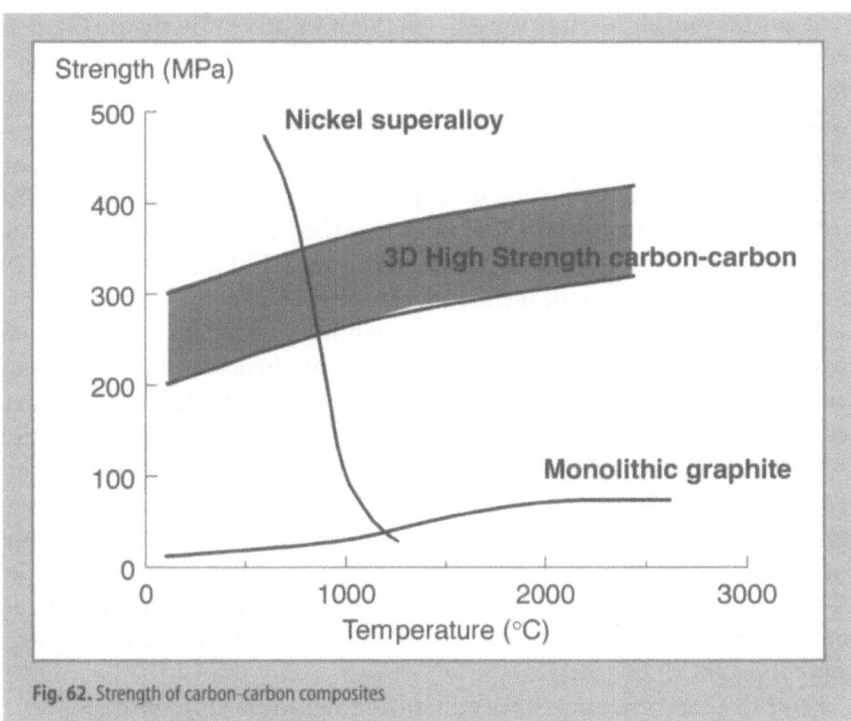

Fig. 62. Strength of carbon-carbon composites

ever, the limited use of carbon-carbon composites in jet engines indicates the difficulty of providing reliable protection.

15.4
Ceramic Matrix Composites

Ceramic matrix composites are candidate materials for high temperature applications in gas turbines, heat exchangers, space vehicles and nuclear reactors [17, 18, 19].

It was noted earlier that, while significant improvements have been made in fracture toughness in several ceramics, lack of reliability and defect tolerance have proved to be major limiting factors in the further application of ceramics in high temperature structural components. The first indications that these problems may be overcome by ceramic matrix composites (CMC's) came from work carried out in the 1960's on the incorporation of carbon fibres in glass [20]. Manufacture involved impregnating fibre tows with fine particles of glass powder suspended in a solvent containing an organic binder. The impregnated tows were filament wound to produce "pre-preg" sheets, which were hot pressed in a graphite die to produce a densified composite. The resulting composites were found to have good strength and toughness. The oxidation behaviour of carbon fibres was a limitation but the availability of the polycarbo-silane derived SiC fibre Nicalon in the early 1980's gave the promise of improved oxidation resistance. This fibre was incorporated in glass ceramic matrices [21, 22]. Combustion panels in this material were trialled in a US military jet engine development programme in 1993.

The need to increase temperature capability over that of glass ceramic matrix composites led to the incorporation of Nicalon fibres in SiC matrices. The manufacturing process involves the deposition of the matrix within the low-density fibre perform by chemical vapour infiltration (CVI) using a gaseous precursor, such as a chlorosilane [23]. An alternative is to infiltrate with a polymer precursor, which is then pyrolysed. Both processes are slow and several impregnation cycles may be necessary to give the required matrix density. Manufacture is fully developed at an industrial level and large parts have been manufactured. Exhaust nozzle flaps were engine tested in the late 1980's and are standardized in some military jet engines.

While SiC/SiC materials have enhanced toughness [24], fibre properties limit strength and temperature capability. In current materials, the Young's Modulus of the Nicalon fibres (190 GPa) is lower than that of the CVI SiC Matrix (400 GPa) and so there is no composite strengthening – Fig. 63. The temperature capability of SiC/SiC is 1100-1200°C – much lower than that of sintered SiC – largely due to fibre characteristics. Nicalon has non-stoichiometric composition – it contains oxygen and excess carbon, which reduces its high temperature properties – Fig. 64 – because of transformation to equilibrium products on exposure at high temperatures.

Another key factor is interface behaviour. As was noted earlier, the properties of the fibre-matrix interface are of prime importance in determining composite behaviour. With interfaces of appropriate strength, fibre debonding, fibre fracture and fibre pull-out act to improve the toughness of composites. In composite systems such as SiC/glass ceramic, a very thin (10–40μm) carbon rich layer is formed at the fibre/matrix interface during processing. This layer is believed to form a bond strong enough for load transfer yet weak enough to debond and allow fibre full out. However, oxidative instability on exposure to high temperatures results in disappearance of the carbon layer and reduced toughness [25].

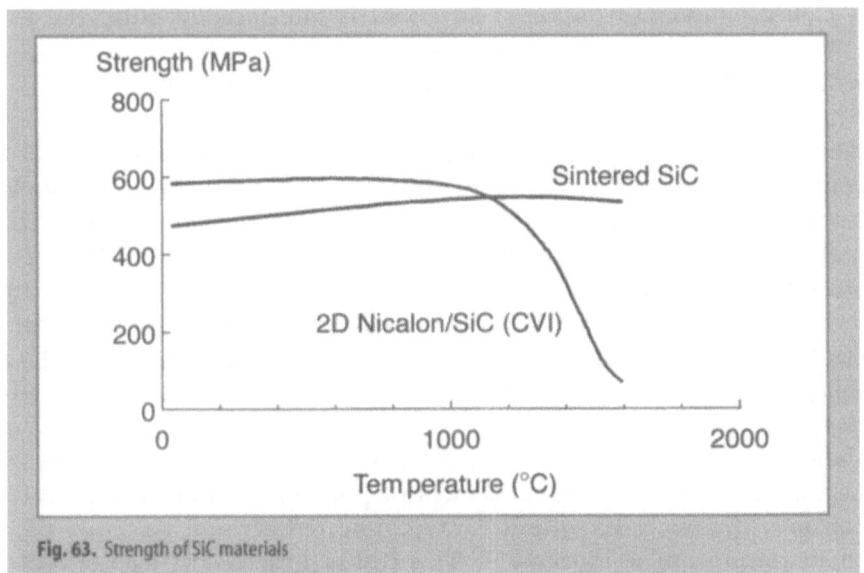

Fig. 63. Strength of SiC materials

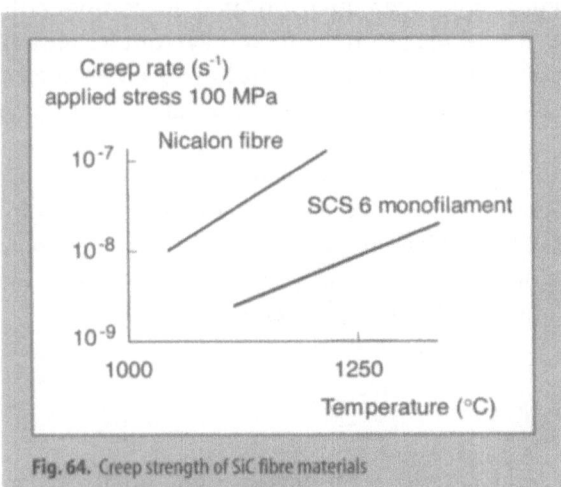

Fig. 64. Creep strength of SiC fibre materials

While some considerable success has been achieved in ceramic composite technologies, further work is necessary. Major problem areas [26, 27, 28] have been defined as:
- Lack of high temperature fibres
- High cost of processing/manufacturing methods

- Control and optimization of fibre/matrix interface, including the development of stable fibre coatings
- Lack of design methodology and life prediction capability

Fibre developments have included polycrystalline SiC fibres produced by continuous sintering and by CVD onto carbon substrates, and single crystal oxide, mixed oxide and eutectic fibres produced by various techniques. Several approaches are being adopted to produce higher temperature fibre coatings and to develop more cost effective manufacturing technology. The directed oxidation of molten metal in the Lanxide process, for example, is claimed to have near-net-shape capability [29, 30]. In this process, the reaction product forms at the surface of the molten metal and growth proceeds with a continuous supply of molten metal via channels in the oxidation reaction product. Doping of the melt is usually necessary to achieve economic rates of process reaction. The product consists of a three dimensional interconnected ceramic structure with small amounts of residual metal. This residual metal may lead to problems and techniques for converting it to a ceramic may be possible. Reinforcements of various types have been incorporated in components. Early work focussed on the Al_2O_3/Al system, with and without Nicalon fibre reinforcement, and many other ceramic systems have been produced. Various automotive engine components have been made and have passed feasibility testing.

The importance of high temperature composite materials to the US DoD/NASA IHPTET program (Integrated High Performance Turbine Engine Technology) [31] is likely to lead to progress in the problem areas. Whether the resulting materials technologies prove to be cost effective, even for aerospace applications, remains to be seen.

15.5
Intermetallic Matrix Composites

The damage tolerance of intermetallic materials can potentially be improved by the incorporation of composite characteristics via intermetallic matrix composites.

Titanium MMC will be limited to a temperature of around 600°C. Above this temperature it will be necessary to use intermetallic matrix composites (IMC) and work has been carried out on both Ti_3Al and TiAl as matrices. TiAl, particularly, has the potential to operate at significantly higher temperatures than TiMMC. Processing temperatures are higher and α mismatch is greater than for TiMMC. Thus interface problems will be more severe for IMC. Preferred systems have not been identified and matrix, interface coating, fibre and manufacturing route decisions have yet to be taken. Foil production, for example, is much more difficult than for conventional Ti alloys.

For applications above 1000°C possible matrices include aluminides such as NiAl, $NbAl_3$ and silicides such as $MoSi_2$. Candidate fibres include SiC, Al_2O_3 and tungsten alloy fibres. As with TiMMC, α mismatch and fibre/matrix reaction control are key issues, more severe than for titanium aluminide matrix composites. Various techniques have been used to produce continuously and discontinuously reinforced composites, from plasma spraying to directional solidification. The concept of producing in-situ composites by the directional solidification of eutectic alloys was investigated in the early 1970's [32]. There has recently been renewed interest in developing intermetallic based DS eutectics for high temperature structural applications, such as NiAl refractory metal composites [33]. It has been claimed that stress rupture properties comparable with the latest single crystal superalloys can be achieved in in-situ refractory metal intermetallic composites (RMIC) [34]. They are referred to as in-situ composites because directional solidification produces a two-phase microstructure containing a significant amount of intermetallic phase in a metallic matrix – such as niobium based solid solution reinforced with silicides. Continuous Al_2O_3 reinforced $MoSi_2$ and NiAl matrices produced by alternative routes have reportedly been under development for nozzle flaps for advanced aero-gas turbine programmes.

16 Coatings for High Temperature Materials

High temperature materials commonly operate in environments involving various combinations of oxidation, corrosion, erosion and wear. Satisfactory performance requires the component to have resistance to these environments as discussed earlier. In some applications the inherent resistance of the base material must be supplemented by a coating in order to ensure the required component life.

Coating processes such as pack cementation to apply chromium and aluminium to steels and weld deposition have been available, initially in relatively crude form, since the 1920's. In 1935, a piston engine in aircraft propulsion application used chromium plating as a means of salvage and wear protection, although this was not a high temperature application. The engine also used weld-applied Stellite antifrettage pads on valve stems and Brightray oxidation resistant inserts on valve seats [1]. Bulk weld deposition of wear resistant surfaces became standard practice in some general engineering applications e.g. excavation machinery. On a smaller scale, metal deposition by TIG welding to build up eroded or corroded areas of gas turbine rotor and stator blades has become widely applied [2, 3] to avoid the need to replace costly components at a stage when their mechanical life has only been partly used. The use of coatings in industrial gas turbines began in the 1960's with chromium coatings on blades [4]. Around the same time, corrosion problems on aero gas turbine blades required the application of aluminide coats, the operating temperatures being too high for chromium coats.

The use of coatings has progressed from the early "problem solving" type of application to the situation in which surface coatings of various types are an integral part of the component design process.

16.1
Corrosion/Oxidation Resistant Coatings

These coatings involve the protective oxides Cr_2O_3 and Al_2O_3 as for base alloys, but they can contain higher amounts of aluminium and/or chromium than are present in base alloys because their compositions are not limited by mechanical property considerations. The protective effect of Cr_2O_3 is limited to around 1000°C due to the formation of volatile CrO_3, while Al_2O_3 is effective to higher temperatures.

Many different processes are used to apply corrosion resistant coatings. Where large areas require protection, processes such as cladding or co-extrusion are commonly used. In early coal fired boilers in power generation, corrosion by the combustion gases and ash particles led to severe fireside metal wastage in carbon steel evaporator tubes [5]. Co-extruded tubes manufactured by extruding a composite billet produces a good metallurgical bond between the inner and outer materials. Tubes with type 310 stainless steel as outer layer on carbon steel tubes gave improved performance [6] and, together with modified type 310 on Essehete 1250 steel, have had extensive usage. Under more severe conditions a combination of 50Cr/50Ni over Incoloy 800 has demonstrated cost benefits to the user. Although 50 Cr/50Ni has excellent corrosion resistance it has low ductility and there have been fabricability problems in manufacture [7]. Reduced chromium content to reduce the potential to alpha-chromium phase formation may be beneficial.

Where the components to be coated are much smaller than evaporator and superheater tubes, several different coating processes are available, depending on the coating to be applied. The major processes include:
• Pack cementation
• Thermal spray
• Physical vapour deposition
• Electroplating

The so-called diffusion coatings are applied by several different techniques using the basic chemical vapour deposition (CVD) principle in which gaseous reactants are transported to the component surface and react chemically with it to form a solid phase containing at least one of the elements present in the gaseous reactants.

Chromised diffusion coatings have been widely used in industrial gas turbine engines since the 1960's. They are limited in use to temperatures of around 900°C, but have been very successful at lower temperatures. Their early use in industrial engines was a consequence of the concurrent need to decrease the chromium content of blade superalloys in order to increase their strength. In gas turbines burning blast furnace gas, the coatings gave protection up to 25000 hours and units running on crude oil achieved running times of 75000–95000 hours [4].

Aluminised coats are by far the most common type of diffusion coating, being used extensively to protect a wide range of aero-, marine and industrial gas turbine components. Pack cementation is the most commonly used process [8]. The components to be coated are packed in a mixture of fine aluminium powder (or pre-alloyed powder), a halide energizer and inert alumina. The pack is heated under inert or reducing conditions in the temperature range 700–1100°C. Aluminium monohalide gas reacts with the base alloy to form nickel or cobalt aluminide, depending on base material. Other phases in the Ni-Al or Co-Al systems may also be present and a heat treatment may be required to form the NiAl or CoAl phases. Variations of the process, in which the component is isolated from the powder, can coat components with internal cooling holes via metal- carrying gaseous species. The thickness of the coating on the exterior surface of gas turbine blades is typically 25–75 μm and the aluminium content can be varied from 25–40% depending on process temperature and heat treatment. Since the coatings are formed by inter-diffusion between the depositing aluminium and the base material, the coating composition will be influenced significantly by the base material. From a corrosion resistant point of view the composition may not be ideal. Nevertheless, major improvement in corrosion resistance has been achieved [9].

The ongoing need to increase coating life has led to the development of improved aluminides. The most significant of these contain platinum [10] which introduces a corrosion resistant platinum aluminide phase and also stabilises the nickel aluminide phase, thus increasing corrosion resistance and life. Such coatings are widely used in both aero and industrial turbines.

In the early 1970's the MCrAlY (M = Co, Ni or both Co and Ni) "overlay" coatings were developed. The claimed advantages were:
- Wider selection of chemical composition possible.
- Superior oxidation or corrosion resistance (depending on composition).
- Superior ductility (depending on composition).

- Improved scale adherence in thermal cycling conditions because of Yttrium addition.

MCrAlY coatings were initially applied by physical vapour deposition, using electron beam evaporation technology [11] but the thermal spray process, low-pressure plasma spraying [12], has become an increasingly competitive process. Recently composite electroplating has been used to apply MCrAlY coatings [13]. In the PVD process, the surfaces of the components are cleaned, moved through a vacuum lock and preheated in the coating chamber. They are then rotated in the vapour cloud above the coating alloy, which is melted by EB heating. After coating, the components are usually glass bead peened and heat treated to minimize defects. Typical turbine aerofoil coating thicknesses are 100–150µm. In low pressure plasma spraying (LPPS), sometimes referred to as vacuum plasma spraying (VPS), the gun sprays molten or semi-molten pre-alloyed powder onto the component surface in a chamber under a low pressure of inert gas. The latest equipment is fully automated, with the gun manipulation controlled robotically. Spraying in low-pressure inert gas restricts oxidation of the powder [14]. It is possible to incorporate, via the pre-alloyed powder, additional elements such as hafnium, platinum and silicon, which improve corrosion/oxidation resistance.

Aluminide and overlay MCrAlY coats are compared schematically in Fig. 65, which shows the greater extent of diffusion in the aluminide coat.

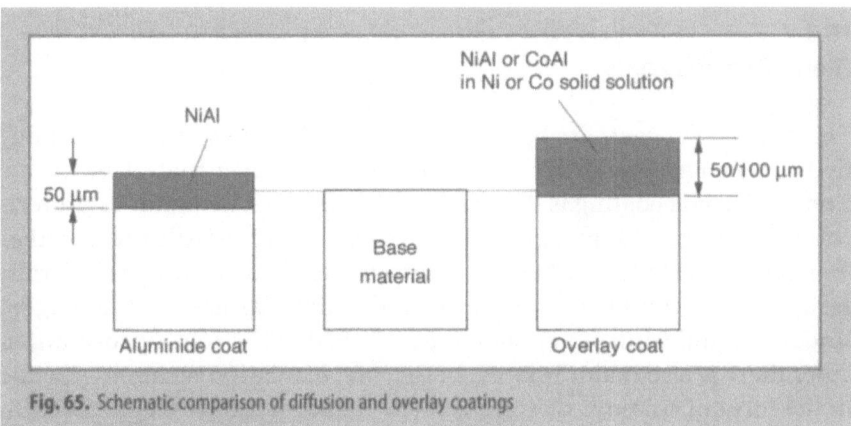

Fig. 65. Schematic comparison of diffusion and overlay coatings

In service, coating degradation occurs by:
- Progressive loss of Cr,Al and other protective elements from the surface by oxidation and corrosion processes
- Coating/substrate interdiffusion, generally resulting in diffusion of Al from the coating into the substrate

The diffusion of aluminum results in a zone at the coating/substrate interface which is depleted in ß-NiAl and the extent of this depleted zone has been used as an indicator of coating condition [15]. Coating life can be extended if aluminum interdiffusion can be suppressed and barrier layers between substrate and coating have been applied with this objective [16].

In addition to the above effects, the coated component will be subjected to mechanical and thermal strains in service. Thermo-mechanical-fatigue testing with test conditions selected to represent service operation is required to determine life, as discussed in chapter 3. However, the inherent strain tolerance i.e. ductility of the coating is composition-dependent and this can be varied more easily in overlay coatings. Aluminide coats have a high ductile-brittle transition temperature (DBTT) even after diffusion heat treatment to reduce aluminium content to 25/30% and this has led to coating cracking on some blades which experienced high cyclic strain range during temperature changes in operation. Before life prediction based on thermal fatigue testing was available, it was possible to improve the ductility of MCrAlY coats by decreasing the Al and Cr content (Fig. 66). This improved the mechanical performance of the blades, albeit at the expense of corrosion resistance [9].

16.2
Thermal Barrier Coats

Thermal barrier coats are based on zirconia, which has a thermal conductivity of around one order of magnitude lower than that of nickel and iron. A zirconia coating is thus an effective thermal insulator to protect a component from a high temperature environment. For this reason thermal barrier coats (TBC) have become standardized coatings on turbine aerofoils and combustors in aero- and industrial gas turbine engines since their initial introduction in the mid 1970's [17]. In reciprocating engines there is also major interest in thermal insulation technology, either in the form of coatings or solid inserts. Its use in diesel engine develop-

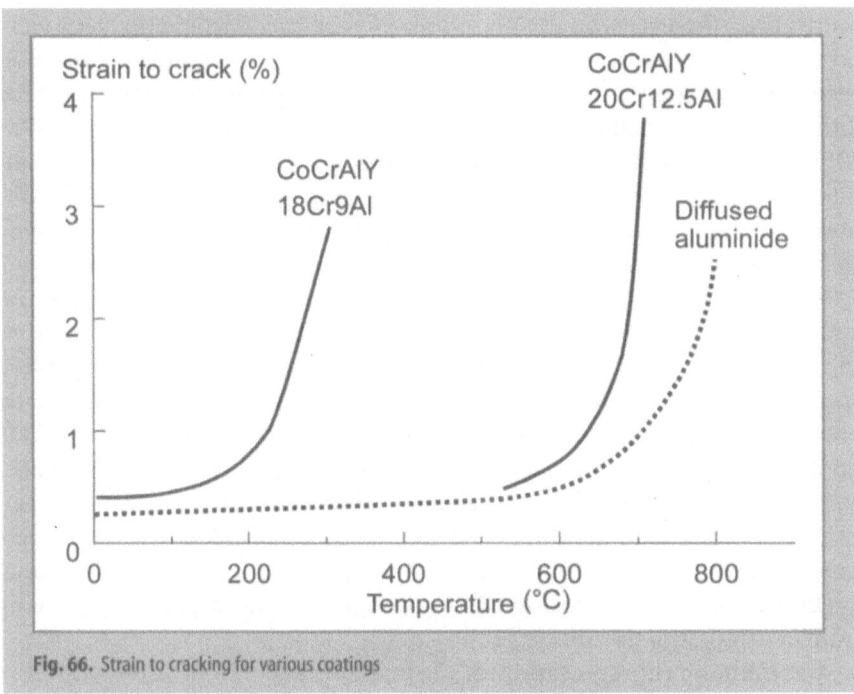

Fig. 66. Strain to cracking for various coatings

ment has reduced heat loss to the cooling system, allowed higher operating temperatures and improved efficiency [18].

As was noted in chapter 14, zirconia is polymorphic. The associated phase changes, and their accompanying volumetric changes, must be avoided to prevent coating spallation. Early coatings used magnesia additions to stabilize the cubic phase. Subsequently, these coats were found to destabilize by reprecipitation of MgO from the ZrO_2 if held for long periods of time at elevated temperature and were thus subject to a maximum operating temperature of about 950°C [19]. The ongoing need to increase the temperature capability of thermal barrier coats led to the development of partially yttria stabilized zirconia (PYSZ) with the addition of around 8% Y_2O_3 to the ZrO_2. This produces a metastable tetragonal phase, which is thermodynamically unstable. However, the reaction kinetics are sufficiently slow that the phase is stable in practice at temperatures above 1200°C, depending on life. Full phase stabilization with the addition of 20% Y_2O_3 is possible, but better performance has been achieved with PYSZ. Bond coats are essential as an interlayer between ceramic and substrate. They bond chemically to the substrate and provide a surface sufficiently rough to bond to the ceramic. Early bond coats were air plasma sprayed Ni-Cr compositions. These had inadequate oxidation resistance to cope with the increased operating temperatures required and were superseded by air sprayed and subsequently argon shrouded or LPPS MCrAlY coats with improved oxidation resistance. In early combustor applications, the TBC reduced metal temperatures by some 50°C and increased the life by a factor of 2–3, with more recent coats showing major further improvement.

Application of TBC's was subsequently extended to turbine rotor and stator blades in order to increase engine efficiency by allowing increase in turbine inlet temperature and/or reduction in air used for air cooling the components. Turbine blade metal temperatures were reduced by 150°C. More recently, improved performance has been claimed for TBC's applied by the electron beam physical vapour deposition (EBPVD) process. The ceramic structure produced by this process consists simplistically of individual, free standing columns, each well bonded to the substrate but essentially free to move relative to adjacent columns during thermal cycling. Thus, the generation of long-range stresses is prevented, giving improved coating durability [20].

A schematic representation of plasma sprayed TBC's is illustrated in Fig. 67.

The benefits of TBC can only reliably and reproducibly be achieved with detailed behavioural understanding and associated life prediction methodology. The basic TBC failure mode is ceramic spalling at the ceramic/bondcoat interface on thermal cycling during engine operation,

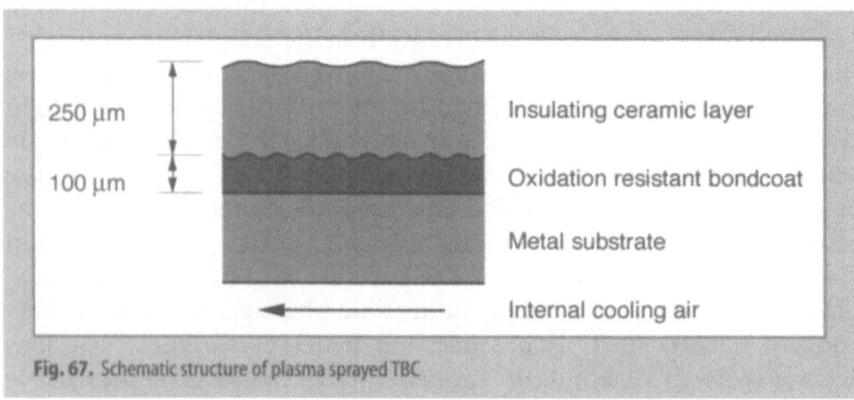

Fig. 67. Schematic structure of plasma sprayed TBC.

In the figure:
- 250 μm — Insulating ceramic layer
- 100 μm — Oxidation resistant bondcoat
- Metal substrate
- Internal cooling air

due largely to the differential thermal expansion of ceramic and metal. Two degradation mechanisms are involved – mechanical damage attributed to cyclic inelastic strain in the ceramic and oxidation of the bondcoat at the ceramic/bondcoat interface [21]. Residual stress and ceramic microstructural control are very important in the performance of plasma sprayed TBC. Relatively dense coatings fail prematurely with typical brittle ceramic failures. Control of the spraying process to incorporate controlled porosity and microcracking results in enhanced ceramic compliance. Residual stress in the coating is controlled by selecting spraying parameters to ensure that the metal substrate temperature is maintained within specified limits.

It has been found that the spallation life of TBC is quantifiably predictable. The life is influenced by the severity of thermal cycling (and hence the severity of cyclic strain in the ceramic) and by the maximum operating temperature which determines the bondcoat oxidation. These factors are included in the behavioural modelling and life prediction methodology [22].

References

Chapter 1:

[1] R. Blum, Materials for Adv. Power Eng. ed., D. Coutsouradis et al, 1994, p. 15, Kluwer Academic Pub. Netherlands

[2] S. Newsam, R&D of High Temp. Materials for Industry ed., E. Bullock, 1989, p. 405, Elsevier Applied Science

[3] M. de Piolenc, Gas Turbine World, Sept/Oct 1992, p. 46

[4] C. Gibb & A.T. Bowden, Jnl of Roy. Soc. Arts. 95, 1947, p. 4739

[5] W. Schlachter & G.H. Gessinger "High Temp. Materials for Power Engineering" ed E. Bachelet et al, 1990, p. 1, Kluwer Academic Pub. The Netherlands

[6] F. Whittle, Proc. Inst. Mech. Eng. 152, 1945, p. 419

[7] G.W. Meetham, The Metallurgist & Materials Technologist, Nov. 1976, p. 589

[8] C.T. Sims, "Superalloys 1984", TMS-AIME, 1984, p. 399

[9] K.C. Barraclough, Metals & Materials, Dec. 1989, p. 707

[10] K. Upadhya, JOM, May 1992, p. 15

[11] L.C. Lindgren et al, SAE Technical Paper 831520, 1983

Chapter 2:

[1] O.C. Zienkiewicz, "The Finite Element Method", Third Edition, Mc. Graw Hill, UK, 1985, p. 135

[2] A. Barker & E. Pinder, Ironmaking & Steelmaking, 1976, p. 195

[3] K.C. Barraclough, Iron & Steel, Sept/Oct 1962, p. 2

[4] F. Turner, "The Development of Gas Turbine Materials", ed G.W. Meetham, 1981, p. 177, Applied Science Publisher Barking, UK

[5] B. Baudelet, Metals & Materials, 1974, p. 117

[6] S. Viswanathan et al JOM Sept 1992 p. 37

[7] D.A. Ford in "The Development of Gas Turbine Materials" – ed G.W. Meetham 1981 p. 147, Applied Science Publishers Barking UK

[8] B. Hicks Ibid p. 229

[9] T.G. Gooch in "R & D of High Temperature Materials for Industry" – ed E. Bullock 1989 p. 289, pub. ElsevierApplied Science London and New York

[10] L.A. Bertram et al JOM 50 (3) 1998 p. 18

[11] D.K. Melgaard et al JOM 50 (3) 1998 p. 13

[12] J.A. Van den Avyle et al JOM 50 (3) 1998 p. 22

[13] P.D. Spilling, Mat Sci Tech 1992 8, p. 145

[14] R.A. Wallis & I.W. Craighead JOM 47 (10) 1995 p. 69

[15] A.L. Purvis et al JOM 46 (1) 1994 p. 38

[16] J.S. Tu et al JOM 47 (10) 1995 p. 64

[17] A.C. Pickard, Mat Sci Tech 1987 3, p. 743

[18] V. Regis et al in "Materials for Advanced Power Engineering" Kluwer Academic Publishers The Netherlands 1994 p. 1779

[19] K.C. Antony & G.W. Goward in "Superalloys 1988" The Metallurgical Society p. 745

[20] Y. Lindblom, Mat Sci Tech 1985 1, p. 636

[21] P. Brauny, Mat Sci Tech 1985 1, p. 719

Chapter 3:

[1] J.R. Nicholls & S.R.J. Saunders, High Temp. Tech. 7 (4), 1989, p. 193

[2] J.A. Conner & W.B. Connor, JOM 46 (12) 1994 p. 35

[3] G.C. Wood & F.H. Stott, Mat. Sci. Tech. 3, 1987, p. 519

[4] G.Y. Lai, JOM, Nov. 1991, p. 54

[5] G. Sorell, NACE publication 58-2, 1957

[6] P. Kofstad, "High Temp. Corrosion" – New York: Elsevier, Applied Science, 1988

[7] J. Stringer, Mat. Sci. Tech. 3, 1987, p. 482

[8] N.S. Bornstein, JOM 48 (11) 1996 p. 37

[9] C. Cain Jr. & W. Nelson, J. Eng. Power Trans ASME, Oct 1961, p. 468

[10] S.K. Bose & H.J. Grabke, Z. Metallk. 69, 1978, p. 8

[11] R.A. Perkins et al, "Properties of High Temperature Alloys", Pennington NJ, Electrochemical Soc., 1976

[12] H.E. Gresham, Metals and Materials, 1969, p. 433

[13] A.C. Jesper, Proc. Inst. Mech. Eng. 180, 1966, p. 265

[14] A.H. Cottrell, "The Mechanical Properties of Matter", Kreiger, Melbourne-Florida, 1964

[15] J. Weertman, TASM 61, 1968, p. 681

[16] O.D. Sherby & J.L. Lytton, Trans. Am. Inst. Min. Metall. Pet. Eng. 206, 1956, p. 928

[17] S.S. Manson, Exp. Mech. 5 (7), 1965, p. 193

[18] C.W. Brown et al, Metal Science 18, 1984, p. 374

[19] D.M. Nissley et al NASA CR 189223, 1992

[20] H.L. Bernstein et al "Thermo-mechanical-fatigue behaviour of materials", ed H. Schitoglu ASTM, 1993, p. 212

[21] T. Baumgärtner et al "High Temperature Materials for Power Engineering", Kluwer Academic Publishers Liège Belgium, 1990, p. 1151

[22] J. Bressers et al in "The Materials Challenge" Sept 1994 p. 4, Joint Research Centre Petten The Netherlands

Chapter 4:

[1] W.I. Mitchell Z. Metallkd. 57, 1966, p. 556

[2] J. Nutting, J.M. Arrowsmith J.I.S.I. London 70, 1961, p. 147

[3] R.F. Dekker, J.R. Mihalisin Trans.Am.Soc.Mat. 62, 1969, p. 481

[4] F. Turner in G.W. Meetham (ed) "The Development of Gas Turbine Materials", Appl.Sci.Pub.Barking UK, 1981, p. 177

[5] R.H. Jeal, Metals & Materials 1985, p. 528

[6] G.J.S. Higginbotham, Mat. Sci. Tech. 2, 1986, p. 442

[7] G.M. Acer & G.H. Meier Oxid. Met. 13, 1979, p. 159

[8] N. Birks et al JOM 46 (12) 1994, p. 42

[9] G.C. Wood & F.H. Stott, Mat. Sci. Tech. 3, 1987, p. 519

[10] T.H. Rhys-Jones, Mat. Sci. Tech., 1988 4, p. 421

[11] L.H. Wolfe, Materials Performance, April 1978, p. 38

[12] M.H. Lewis et al in "Non-oxide Technical & Engineering Ceramics" ed Hampshire (Elsevier 1986) p. 175

[13] D.P. Hasselman, Bull.Am.Ceram.Soc. 49, 1970, p. 1033

[14] M.H. Lewis & G. Leng-Ward, Metals & Materials, 1991, p. 356

[15] R.A.J. Sambell et al J. Mat. Sci. 7, 1972, p. 676

Chapter 5:

[1] D.A. Oliver et al in W.E. Benbow, "Steels in Modern Industry", Iliffe & Sons London, 1951

[2] H. Hochmann, Rev. Met. 48, 1951, p. 734

[3] R.F. Steigerwald et al, The Metallurgist & Materials Technologist, April 1978, p. 181
[4] C.R. Austin et al, Trans.Am.Inst.Min.Metal.Pet.Eng. 162, 1945, p. 84
[5] E.W. Colbeck & J.R. Rait, I.S.I. London, Spec. Report, No 43, 1952
[6] G. Wood & J.R. Rait, Iron & Coal Trades Review, No 481, Feb. 1949, p. 2
[7] C.J. McMahon Jr. et al, EPRI Report NP 1501, Sept. 1980
[8] R.I. Jaffee, Met. Trans. A, 17A, 1986, p. 755
[9] H.W. Kirkby, Proc. Inst. Mech. Eng 180, 1965-1966, p. 1149
[10] K.J. Irvine et al, J.I.S.I. 195, 1960, p. 386
[11] J.H. Woodhead & A.G. Quarrell, JISI, June 1965, p. 605
[12] R.D. Townsend in "R & D of High Temperature Materials for Industry", ed. E. Bullock Pub Elsevier Applied Science, 1989,p. 13
[13] O. Wachter et al, Report of the Research Centre Jülich, Jül-3074, 1995, ISSN 0944-2952
[14] R. Blum, "Materials for Advanced Power Engineering", 1994, p. 15, Kluwer Academic Publishers, The Netherlands
[15] J.E. Truman, The Metallurgist & Materials Technologist, Feb. 1980, p. 75
[16] J.J. Jones in "R&D of High Temp Materials for Industry" ed E. Bullock, Elsevier Applied Science – London, 1989, p. 31
[17] H. Lewis, British Corrosion Jnl. 3, 1968, p. 166
[18] A. Vinter & L.G. Wilbers, JOM 22 (5), 1970, p. 46
[19] L.J. Hull, Metal Progress, Dec. 1959, p. 76
[20] K.J. Irvine et al, JISI, July 1959, p. 218

Chapter 6:

[1] R.J. Greene & F.G. Sefing, Corrosion, 1955 11, p. 315
[2] G.W. Form & J.F. Wallace, AFS Trans 70, 1962
[3] C. Van der Ben, Alloy Metals Rev. 8, Dec. 1950, p. 2
[4] Nickel Development Institute Publication No. 4077, 1967
[5] J.W. Grant & J.C. Morrison, The British Foundryman, May 1971, p. 172
[6] K. Rohrig, Castings Buyer 10(3), 1998, p. 14
[7] J.C. Morrison, Castings Buyer, July 1990
[8] R. Covert et al Nickel Development Institute Publication No. 11018, 1998

Chapter 7:

[1] P.B. Wallis in "The Nimonic Alloys" ed W. Betteridge & J. Heslop, Edward Arnold, London, 1974, p. 459

[2] G.W. Meetham, The Metallurgist and Materials Technologist, 1982, p. 387

[3] W. Betteridge & J. Heslop, "The Nimonic Alloys", 1974, Arnold, London

[4] D.A. Ford, "The Development of Gas Turbine Materials" ed G.W. Meetham, 1981, p. 147, Applied Science Publishers Barking UK

[5] S.T. Wlodek, TASM 57, 1964, p.110

[6] C.H. Lund et al 1969 German Offen, 1921359

[7] P.S. Kotval et al, Met. Trans. 3, 1972, p. 453

[8] S.W.K. Shaw, Metal Progress, March 1979, p. 47

[9] G.B. Thomas & T.B. Gibbons, Metals Tech., March 1979, p. 95

[10] G.W. Meetham, Metals Technology 11, 1984, p. 414

[11] F. Turner in "The Development of Gas Turbine Materials" ed G.W. Meetham, 1981, p. 177, Applied Science Publishers Barking UK

[12] J.G. Smeggil et al Met. Trans. 16A, 1985, p. 1164

[13] R.T. McVay et al "Superalloys 1992" ed S.D. Antolovich et al, 1992, The Minerals, Metals and Materials Soc.

[14] D. Furrer & H. Fecht, JOM 51 (1), 1999 p. 14

[15] R.M. Forbes Jones & L.A. Jackman, JOM 51 (1) 1999 p. 27

[16] P.D. Spilling Mat. Sci. Tech. 1992 8 p. 145

[17] G.H. Gessinger & M.J. Bomford, Int. Met. Revs. No 181, 1974, p. 51

[18] G.I. Friedman & G.S. Ansell, "The Superalloys" ed Sims & Hagel, 1972, p. 427, Wiley and Sons

[19] P. Wightman & E. Hengsberger, Metals & Materials, 1991, p. 676

[20] M. McLean, "DirectionallySolidified Materials for High Temperature Service", 1983, Pub. The Metals Soc. London

[21] G.J.S. Higginbotham, Mat.Sci.Tech. 2, 1986, p. 442

[22] M.J. Goulette et al, "Superalloys 1984", TMS-AIME, 1984, p. 117

[23] K.P.L. Fullager et al ASME 94-GT-169 1994

[24] G.L. Erickson JOM 47 (4) 1995 p. 36

[25] T.M. Pollock et al "Superalloys 1992" ed S.D. Antolovich et al 1992 p. 125 Pub. The Minerals, Metals and Materials Soc.

[26] F.J. Anders et al, Metal Progress, Dec. 1962, p. 88

[27] J.S. Benjamin, Met. Trans 1, 1970, p. 2943

[28] M.G. McKimpson & D. O'Donnell JOM 46 (7) 1994 p. 49

Chapter 8:

[1] F.R. Morral, et al ASM Met. Eng. Q. 1969 9 (2) p. 1
[2] J.Stringer, Mat. Sci. and Tech., 1987, 3, p. 482
[3] J.A. Vaccari, Materials Engineering May 1969 p. 21
[4] H.L. Wheaton, Cobalt 1965 (29) p. 16
[5] A.M. Hall & F.R. Morral, Materials in Design Engineering Feb. 1966

Chapter 9:

[1] J.G. Heyes & R.G.R. Sellors, Metals & Materials, Feb. 1992, p. 86
[2] N.E. Promisel ed, "Science & Technology of Tungsten, Tantalum, Molybdenum, Niobium and their Alloys", 1964, Pergamon Press, London
[3] G.D. Mc.Adam, J.Inst.Met. 93, 1964, p. 559
[4] J.E. Restall, Metals & Materials 1, 1967, p. 241
[5] D.J. Jones, Metall.Mater.Tech. 5, 1973, p. 503
[6] J.L. Briggs & R.Q. Barr, High Temp.-High Press 3, 1971, p. 363
[7] W. Fairhurst, Metall.Mater.Tech. 6, 1974, p. 68
[8] F. Turner in "The Development of Gas Turbine Materials" ed G.W. Meetham, 1981, p. 184, Applied Science Publishers Barking UK
[9] G. Llewelyn, Proc.Inst.Mech.Eng. 180 (3D), 1965-1966, p. 90
[10] J.E. Restall, J. Less Common Metals 16, 1968, p. 11

Chapter 10:

[1] T.W. Farthing, Proc.Inst.Mech.Eng. 191, 1977, p. 59
[2] A. Mitchell, JOM 49, 1997, p. 40
[3] K. Rudinger & D. Fisher, Titanium '80 2 ed Kimura & Izumi pub Warrendale PA TMS, 1980, p. 1907
[4] R.E. Goosey, Iron & Steel Inst.Sp.Pub. 124, 1969, p. 75
[5] M.R. Winstone, J. Less Common Metals 39, 1975, p. 205
[6] J.A. Hall et al Mat. Sci. Eng. 9, 1972, p. 197
[7] H.W. Rosenberg, The Science, Technology & Application of Titanium, 1970, p. 851, pub Pergamon Press Oxford UK.
[8] P.A. Blenkinsop et al Titanium & Titanium Alloys, 1976 pub Plenum Press
[9] D.P. Davies, Metals & Materials, July 1988, p. 417
[10] R.E. Goosey, Metals & Materials, Aug 1989, p. 451
[11] P.J. Bania, JOM 46 (7), 1994, p. 16

[12] R.R. Boyer, JOM 46 (7), 1994, p. 20
[13] D.J. Smith, Metals & Materials, Feb. 1988, p. 79
[14] S.L. Semiatin et al JOM 49 (6), 1997, p. 33
[15] D.M. Ward, Metals & Materials, Sept. 1986, p. 560
[16] D. Driver, Metals & Materials, Aug. 1988, p. 493
[17] J. Klepeisz & S. Veeck, JOM 47 (11), 1997, p. 15

Chapter 11:

[1] D.M. Dimiduk et al, Mat. Sci. Tech. 8, 1992, Vol 8, p. 367
[2] J.C. Chesnutt & J.C. Williams, Metals and Materials, 1990, p. 509
[3] S.J. Balsone, "Oxidation of High Temp. Intermetallics", ed T. Grobstein & I. Doychak, 1989, p. 219, TMS
[4] Y.W. Kim & F.H. Froes, High Temperature Aluminides & Intermetallics ed S.H. Whang et al 1990, p. 465, pub Warrendale PA TMS
[5] M. Yamaguchi, Mat. Sci. Tech. 8, 1992, p. 299
[6] Y.W. Kim, JOM 46 (7), 1994, p. 30
[7] A. Rahmel et al Materials & Corrosion 46, 1995, p.281
[8] S. Becker et al Oxidation of Metals 38 (5/6), 1992, p. 425
[9] M.P. Brady et al JOM 48 (11), 1996, p. 46
[10] P. Bertolatta et al JOM 49 (5), 1997, p. 48
[11] Y.W. Kim, JOM 47 (7), 1995, p. 38
[12] D.P. Pope & S.S. Ezz, Int. Met. Rev. 29, 1984, p. 136
[13] C.T. Liu et al Acta Metall. 33, 1985, p. 213
[14] C.T. Liu & K.S. Kumar, JOM 45 (5), 1993, p. 38
[15] F.H. Froes, JOM, Sept. 1989, p. 6
[16] E.A. Feest, & J.H. Tweed, Mat. Sci. Tech. 8, 1992, p. 308
[17] C.T. Liu et al Scr. Metall. 23, 1989, p. 875
[18] C.G. McKamey et al J. Mater. Res. 6, 1991, p. 1779
[19] P.R. Subramanian et al JOM 48 (1), 1996
[20] K.S. Kumar & C.T. Liu, JOM 45 (6), 1993
[21] G.H. Meier & F.S. Petit Mat. Sci. Eng. A 153 1992 p. 548
[22] M.J. Malony & R.J. Hecht, Mat. Sci. Eng., A 155, 1992, p. 19

Chapter 12:

[1] K.C. Barraclough, Metals & Materials, Dec. 1989, p. 707
[2] R.V. Hillery, J. Vac. Sci. Tech. A4 (6), Nov/Dec 1986, p.2624
[3] F.J. Honey et al, J.Vac. Sci. Tech. A4 (6), 1986, p. 2593
[4] H. Herman, Adv. Mat. Proc. 137 (4), 1990, p. 41

[5] T.N. Rhys-Jones, Surface and Coatings Technology 43/44, 1990, p. 402, Elsevier Sequoia The Netherlands
[6] A.E. Weatherill & B.J. Gill, Met. Mater. 4, 1988, p. 551
[7] M.G. Nicholas & K.T. Scott, Surf. J. 12 (1), 1982, p. 2

Chapter 13:

[1] J.H. Chesters "Refractories-production & properties" I.S.I. London 1973
[2] American Concrete Institute Report A.C.I. 547R-79
[3] G.H. Kesler in I.E. Campbell, E.M. Sherwood (eds), "High Temperature Materials and Technology", J. Wiley&Sons NY 1969, p. 693
[4] L.J. Korb et al, Bull.Am.Ceram.Soc. 60 (11) 1981, p. 1188

Chapter 14:

[1] D.W. Richerson, "Modern Ceramic Engineering", Marcel Dekker, NY 1992
[2] D.B. Marshall & J.E. Ritter et al, Ceram. Bull. 66 (2), 1987, p. 309
[3] A.G. Evans, Mater. Sci. Eng. 71, 1985, p. 3
[4] J.E. Ritter et al, J. Mater. Sci. 15, 1980, p. 2275
[5] D.B. Marshall & A.G. Evans, Acta. Metall. 33 (11), 1985, p. 2013
[6] E.P. Butler, Mat. Sci. Tech. 1, 1985, p. 417
[7] N. Claussen J. Am.Ceram.Soc. 61, 1978, p. 85
[8] S.M. Johnson & D.J. Rowcliffe, SRI International Report to EPRI "Ceramics for Electric Power Generating Systems", Jan 1986
[9] S.J. Dapkunas, Ceram. Bull. 67 (2), 1988, p. 388
[10] J.E. Oakey et al in Materials for Advanced Power Engineering Part II, 1994, p. 1453 ed D. Coutsouradis et al Kluwer Academic Pub. Netherlands
[11] R.A. Newby & R.L. Bannister, Modern Power Systems, Sept. 1993, p. 51
[12] F.F. Lange, Ceram.Bull. 62, 1983, p. 1369
[13] M.H. Lewis & G. Leng-Ward, Metals&Materials, 1991, p. 356
[14] H.E. Helms & F.N. Heitmann, ASME 84-GT-81
[15] J.R. Kidwell et al, ASME 84-GT-166
[16] D.W. Richerson & K.M. Johansen, Final Report DARPA/Navy Contract N 00024-76-C-5352, May 1982
[17] G. Rogers et al SAE Technical Paper No. 900452, 1990, Warrendale PA

[18] C.G.E. Mangin et al JOM 45 (6), 1993, p. 23
[19] S.D. Slookey, Ind.Eng.Chem. 51(7), 1959, p. 805
[20] C.A. Fucinari & V.D.N. Rao, "Ceramic Regenerator Systems Development" programme, NASA CR-159707 NASA-Contract DEN 3-8 Oct 1979

Chapter 15:

[1] J.D. Birchall et al in "Strong Fibres" ed A. Kelly & S.T. Mileiko, N.Holland, 1983
[2] E..A. Feest, Metals and Materials, 1988, p. 273
[3] R.L. Trumper Metals & Materials, Nov 1987, p. 662
[4] W. Wei, Metals and Materials, Aug.1992, p. 430
[5] P.R. Smith & F.A. Froes, JOM 36 (3), 1984, p. 19
[6] D. Upadhyaya et al JOM 46 (11), 1994, p. 62
[7] R.A. MacKay et al JOM 43 (5), 1991, p. 23
[8] P.G. Partridge & C.M. Ward-Close, Inter. Mater. Rev. 38 (1), 1993, p. 1
[9] F.E. Wawner et al SAMPE Quarterly 14 (3), April 1983, p. 39
[10] G.A. Owens, Proc. Int. Conf. On Composite Materials, 1988, p. 747 ed I.C. Visconti, I.C. Pub. Coop. Univ. Nap.
[11] S.W. Kandebo, Aviation Week & Space Technology, Aug. 22.1994, p. 21
[12] M.S. Wright in IE Campbell (ed), "High Temp Technology", J.Wiley & Sons, New York, 1956, p. 2
[13] J.D. Buckley, Ceram. Bull. 67 [2], 1988, p. 364
[14] F.K. Ko, Ceram. Bull. 68 [2], 1989, p. 401
[15] G. Savage, Metals and Materials, 1988, p. 544
[16] J.R Strife & J.E. Sheehan, Ceram. Bull., 67 [2], 1988, p. 369
[17] D. Ronby & P. Reynaud, Compos. Sci. Tech. 48, 1993, p. 109
[18] P.K. Liaw et al Acta Met. Et Mater. 44, 1996, p. 2101
[19] P.K. Liaw et al J. Nucl. Mater. 219, 1995, p. 93
[20] R.A.J. Sambell et al, J.Mat. Sci 7, 1972, p. 676
[21] K.M. Prewo & J.J. Brennan, J.Mat.Sci. ,17, 1982, p. 1201
[22] K.M. Prewo et al, Ceram.Bull., 65 [2], 1986, p. 305
[23] D.P. Stinton et al, Ceram.Bull., 65 [2], 1986, p. 347
[24] P.J. Lamicq et al, Am.Ceram.Soc.Bull., 65 [2], 1986, p. 336
[25] R.J. Kerans et al, Ceram.Bull., 68 [2], 1989, p. 429

[26] C.Y. Ho & S.K. El-Rahaiby, Ceramic Engineering & Science Proceedings (Am.Ceram.Soc.), July/Aug 1992, p. 3
[27] J.J. Mecholsky Jr., Ceram. Bull. 68 (2), 1989, p. 367
[28] E.L. Courtright, Ceram. Eng. Sci. Proc. 12 (9-10), 1991, p. 1725
[29] B.W. Sorenson et al Turbomachinery International, Sept/Oct 1990, p. 20
[30] M.S. Newkirk, Ceram. Eng. Sci. Proc. 8 (7-8), 1987, p. 879
[31] N.M. Tallon, Ceram.Eng.Sci.Proc., 12 [7-8], 1991, p. 957
[32] H.E. Cline & J.L. Walter, Met. Trans. 1, 1970, p. 2091
[33] K.S. Kumar & G. Bao, Comp. Sci. & Tech. 52, 1994, p. 127
[34] M.R. Jackson et al JOM 48 (1), 1996, p. 39

Chapter 16:

[1] D.F. Bettridge, "Surface Engineering and Heat Treatment" ed., P.H. Morton, Institute of Metals, London, 1991, p. 43
[2] K.C. Anthony & G.W. Goward, "Superalloys 1988" ed., S. Reichman et al, The Metallurgical Soc., 1988, p. 745
[3] P. Brauny et al, Mat. Sci. Tech. 1, 1985, p. 719
[4] R. Bürgel, Mat. Sci. Tech. 2, 1986, p. 302
[5] A.J.B. Cutler et al, Metallurgist & Materials Technologist, Feb. 1981, p. 96
[6] D.B. Meadowcroft, Met. Sci. Eng. 88, 1987, p. 313
[7] T. Flatley & C.W. Morris in "R&D of High Temp. Materials for Industry" ed., E. Bullock, Elsevier Applied Science, 1989, p. 129
[8] J.E. Restall in "The Development of Gas Turbine Materials" ed., G.W. Meetham, Applied Science Publishers, 1981, p. 259
[9] G.W. Meetham, Mat. Sci. Tech. 2, 1986, p. 290
[10] G. Lehnert & H.W. Meinhardt, UK Patent No 1282530, 1970
[11] D.H. Boone, Mat. Sci. Tech. 2, 1986, p. 220
[12] A.R. Nicholl et al, Surf. Engng. 1, 1985, p. 59
[13] F.J. Honey et al, J.Vac. Sci. Tech. A 4, No 6, 1986, p. 2593
[14] A.R. Nicholl et al, Mat. Sci. Tech. 2, 1986, p. 214
[15] N.S. Cheruvu et al Jnl. of Metals, 1996, p. 34
[16] R.A. Page & G.R. Leverant, Jnl. of Eng. for Gas Turbines & Power 121, April 1999, p. 313
[17] A. Bennett, Brit. Ceram. Soc. Proc. No 34, 1984, p. 207
[18] R. Kamo et al, NTIS Conference (790747), 1979

[19] A. Bennett, Mat. Sci. Tech. 2, 1986, p. 257

[20] S.M. Meier & D.K. Gupta, TASME 116, 1994, p. 250

[21] T.A. Cruse et al, Jnl. of Eng. for Gas Turbines & Power 110, Oct. 1988,
      p. 610

[22] J.T. DeMasi-Marcin et al, Jnl. of Eng. for Gas Turbines & Power 112,
      Oct. 1990, p. 521

Index